Plants and Human Conflict

Plants and Human Conflict

By

Eran Pichersky

CRC Press is an imprint of the
Taylor & Francis Group, an **informa** business

CRC Press
Taylor & Francis Group
6000 Broken Sound Parkway NW, Suite 300
Boca Raton, FL 33487-2742

Printed on acid-free paper

International Standard Book Number-13: 978-1-138-61530-4 (Paperback)
International Standard Book Number-13: 978-1-138-61531-1 (Hardback)

Library of Congress Cataloging-in-Publication Data

Names: Pichersky, Eran, author.
Title: Plants and human conflict / author: Eran Pichersky.
Description: Boca Raton, FL : CRC Press, Taylor & Francis Group, 2018. |
Includes index.
Identifiers: LCCN 2018016698| ISBN 9781138615304 (pbk. : alk. paper) | ISBN
9781138615311 (hardback : alk. paper)
Subjects: LCSH: Plants and civilization. | Plants and history. | War.
Classification: LCC SB107 .P543 2018 | DDC 630--dc23
LC record available at https://lccn.loc.gov/2018016698

Visit the Taylor & Francis Web site at
www.taylorandfrancis.com

and the CRC Press Web site at
www.crcpress.com

For if someone were to assign to every person in the world the task of selecting the best of all customs, each one, after a thorough consideration, would choose those of his own people, so strongly do humans believe that their own customs are the best ones. Therefore, only a madman would treat such things [i.e., people's customs] as a laughing matter.

(Herodotus, *The Histories* 3.38, 5th century BC)

"I can't believe that!" said Alice.

"Can't you?" the Queen said in a pitying tone. "Try again: draw a long breath, and shut your eyes."

Alice laughed. "There's no use trying," she said: "one can't believe impossible things."

"I daresay you haven't had much practice," said the Queen. "When I was your age, I always did it for half-an-hour a day. Why, sometimes I've believed as many as six impossible things before breakfast."

(Lewis Carroll, *Through the Looking Glass*, Chapter 5: Wool and Water. Composed in 1858–1862 during and around the time of the Second Opium War, published in 1871)

Contents

Preface

Humans naturally relate to other animals more than they do to plants. Those of us who have taken formal biology classes know, at least on the intellectual level, that plants are living organisms just like animals are. But the majority of us now grow up in an urban environment and think about plants mostly as food, or as more or less inanimate objects in the landscape.

There may be many reasons why people make a clear distinction between plants and animals, but I think it is fair to say that the majority of lay people consider the capacity to move to be the most important indicator of life. The visual sense is the dominant sense in the human species, and since animals clearly move around, they catch our attention. And we believe that history – events – can only be created by things that move around and interact with each other.

I grew up in Israel, a country located on the eastern shores of the Mediterranean Sea. The area has been repeatedly impacted by major historical events for thousands of years, often being the locale where major powers, from the ancient Egyptian kingdoms to the Babylonian, Assyrian, Persian, Greek, Roman, Byzantine, and Arab empires, and a few others, met each other – sometimes peacefully, but more often clashing violently. The modern State of Israel itself has had its share of violent conflicts over its relatively short time of existence, so it is not surprising that since my early life I have developed an interest in history in general and war history in particular.

My interest in plants and their influence on human affairs developed more slowly. Until the age of five, I lived on a communal farm, known as a Kibbutz, surrounded by wheat fields and orchards. The Kibbutz was situated right on the border with the territory called the West Bank, then controlled by Jordan. Between the two countries was a one- to two-mile wide no man's land of perfectly good agricultural land that was going to waste because, under the existing ceasefire agreement, the residents of neither Israel nor Jordan were allowed to cultivate it. When my family and I moved to the city, I used to take every opportunity I had, mostly during school vacations, to go back to the farm where some of my relatives still lived. Nonetheless, until I was 18 years old I generally led a typical life of a city boy in Israel in those days – going to school and, after school, playing mostly in the streets. There were very few plants in my daily life in those days.

All this changed when, after graduating from high school, I was conscripted into military service. Fortunately, I was given some choice in the kind of military unit that I could join, and I decided to volunteer to an army unit that trained soldiers for ground combat, mostly in small groups. Israel has a relatively warm climate, with temperatures rarely dipping below the freezing point, so plants both wild and cultivated grow year-round as long as water is available. Israel does have a diversity of microclimates, ranging from alpine to desert, and with a corresponding diversity of native flora and fauna. My military training – and anyone who has served in the military knows that training, rather than fighting, is the major activity throughout one's service – involved being out in the country day and night, seven days a week. We did a lot of marching from point A to point B, never using paved roads but rather walking through cultivated fields, natural brush, forest, and desert (which, unlike the general impression of being barren, has a very rich and distinct flora as well as fauna). Once we arrived at our destination, we would practice attacking an imaginary enemy, which required us to intermittently crawl on the ground and then race from one natural shelter to another, all the while stepping on plants and hiding behind plants. Sometimes we actually pretended to be plants by camouflaging ourselves with uprooted vegetation or tree branches attached to our clothes and our helmets.

I have strong visual memories of the many plants I encountered during the years I spent in the army. But my olfactory memories of this time are much, much stronger. I can still recall the distinct smells of plants permeating the air wherever I went, and the smell of the crushed plants as I stepped or crawled on them. Smell is the sense with the strongest emotional "pull." I may not have been fully aware of it at the time, but in retrospect I think that my close encounters with plants during my army years did lead me to change my original idea of studying mathematics in college and instead pursue a life-long career of studying plants.

During this career, which began at universities in California, then in New York, and, for the last 30 years or so, in Michigan, I have carried out research on the genetic and biochemical basis for the synthesis of many different compounds in plants, including odorous ones. As will be detailed in the various chapters of the book, plants have evolved the ability to synthesize many more compounds than animals do, and these chemicals have served many functions in the plant. They may be basic building blocks of the plant cells and bodies, but they may also be scent and pigment chemicals that are found in flowers and fruit and thus attract animals to visit the flowers and pollinate them or eat the fruit and disperse the seeds away from the mother plant. Moreover, the twin facts that plants produce many compounds that animals cannot, yet animals need these chemicals for their own metabolism, make plants the targets of unwanted attention by animals that try and consume them. It is therefore not surprising that natural evolution has led to plants developing the ability to also produce many toxic compounds that serve as a defense against animals (and fungi and bacteria, too).

As I studied how plants make all these different compounds for their own needs, I became aware from reading general articles about such compounds that many of them have also been highly desired commodities for people. Some are of course the basic staples – starch, sugar, oils – as well as additional essential nutrients such as vitamins that humans, like other animals, cannot make and therefore need to obtain from plants in order to survive. But beyond the plant compounds that are necessary for subsistence, humans have made extensive use of plant material. Until recently, the vast majority of medicinal compounds came from plants, and even today the majority of our medicines are either plant chemicals or modified plant chemicals, such as aspirin and various opioid pain relievers. In addition, there are also plant-derived compounds that are used for making various items that are essential for advanced human society and culture. In this category, one finds rubber, which is today used for more than 40,000 different human artifacts, and cellulose, which in its various manifestations is the basis for, among others, clothing, paper, and wood for house construction and furniture. Then there are many plant compounds, such as perfumes and flavors, and mind-altering substances, such as caffeine and alcohol, that contribute to making our lives more fun and pleasant.

What I also learned from pursuing my amateur interest in history is that plant chemicals have often been involved in violent human conflicts. This is not surprising: the tendency of people to try to obtain what they need and want by force is well known. I should also quickly add that I was certainly not the first one to realize it – far from it. In fact, many excellent books have been published about specific plants that have been important to people at one time or another. Rubber, tea and coffee trees, cotton and spice plants – these and other plants have been celebrated for their contribution to human affairs, both for progress during peacetime and for their involvement in human warfare and conflict, and many of these studies are cited in this book.

While I enjoyed reading books about how specific plants played a role in human history, I felt that, somehow, looking at one plant at a time misses an essential point. Moreover, and this second point is really linked to the first, these books were typically written by historians, not plant biologists, so that the actual attributes that made these plants essential to humans were described in general terms, without much consideration of how such attributes evolved in the plants in the first place and how they served the plants themselves. In other words, humans were always in the center of the story, using this plant or that, but the plants themselves were completely passive actors – not really alive.

The more I thought about it, and the more I considered the lives of plants and the lives of humans, separately and as they intersected, I began to appreciate that placing humans at the center of the story is no more correct than placing planet Earth at the center of the universe. It is an understandable perspective for humans to employ, but one that ultimately misses the bigger picture.

To see the bigger picture, and to be able to deduce some basic rules that govern the interactions of plants and humans and the impacts of such interactions on both plants and humans, I think it is necessary to combine the history of humans with the histories of the many plants that we have interacted with, all in one story frame. Furthermore, combining these histories will also require combining the biological and historical viewpoints. This is what I tried to achieve in this book, and, being a biologist by training, as I gathered the histories of different plant species' interactions with humans I tried to infuse them with as much biological context as I found necessary and useful.

A complete description of shared human–plant history would be an enormous task, one that would be very difficult to achieve in a single book. The vastness of the topic might in fact make it more difficult to uncover the essential rules that govern such interactions. After much internal deliberation, therefore, I chose to distill this task by focusing on the involvement of plants specifically in human warfare and violent conflicts in general. I did so because I believe that it is during such events that the biological features of both plants and humans are most evidently in play. This book therefore recounts some of the most horrendous, albeit fascinating, chapters in human history, many of which readers may be familiar with. But these events are recounted in somewhat unfamiliar ways, by putting plants at the center of the story, or at least co-equal to the human actors. My goal in doing so was to achieve for myself, and to convey to the readers, a better understanding of both plant and human history. It is my hope that after reading this book, the reader will come away with the newly acquired recognition and appreciation that there is no human biology and history without plant biology and history, and, at least for the domesticated plant species discussed in this book, the opposite is true as well.

Acknowledgments

I would like to thank Liza Pichersky, Anthony Cashmore, Charles Yocum, Laura Olsen, and Noam Greenspoon for reading all or parts of the manuscript and providing many useful comments on style and substance. I would especially like to thank Mayumi Matsuba for making the beautiful plant drawings, Dale Austin for the expertly drawn maps, and Liza Pichersky for photography. Finally, I thankfully acknowledge the John Simon Guggenheim Memorial Foundation for a fellowship award that for a year freed me from all other academic obligations to concentrate on the research and writing of this book.

Author

Eran Pichersky is the Michael M. Martin Collegiate Professor in the Department of Molecular, Cellular, and Developmental Biology (MCDB) at the University of Michigan. He received his B.Sc. degree from the University of California, Berkeley in 1980, and a Ph.D. from the University of California, Davis in 1984. After doing research as a postdoctoral fellow at the Rockefeller University in New York, he has been on the faculty of the University of Michigan since 1986, serving as the first Chair of the newly created MCDB Department from 2001–2003. His awards include a Fulbright Fellowship and an Alexander von Humboldt Fellowship, both received in 2000, and a Guggenheim Fellowship in 2015. He was elected a Fellow by the American Association for the Advancement of Science (AAAS) in 2012 and by the American Society of Plant Biologists in 2017. Dr. Pichersky has served on the editorial boards of several major scientific journals that cover plant research and edited (with Dr. Natalia Dudareva) the book *Biology of Floral Scent* (CRC Press, 2006).

Dr. Pichersky's research has concentrated on identifying the myriad compounds that are found uniquely in plants, many of which are extensively used by people, with emphasis on those that impart scent and flavor. His group further elucidates how plants synthesize these compounds, and how this information can be used to enhance the production by plants of such valuable chemicals. Over the years, Dr. Pichersky's research group has collaborated with many other research groups around the world, and Dr. Pichersky himself has spent extensive time as a visiting scholar doing research at scientific institutes around the world, including the United States, Germany, Israel, and Australia. Dr. Pichersky has authored approximately 250 reports, reviews, letters, and editorials in scientific publications, and is a recipient of several patents.

1 Natural Resources as Causes of Violent Conflicts

UNIVERSAL HISTORY

The term "history" generally refers to the field of study of past events that have involved humans. It is often restricted to the period after writing appeared and written records could be found to provide proof of the occurrence of specific events (although clearly many documents, past and present, are not truthful). Events that occurred before the invention of writing are sometimes classified as pre-history. Such ancient events can nevertheless be reconstructed from archeological evidence – human remains and the remains of human artifacts. But history is more than just uncovering past events. Serious students of history harbor the desire to understand why historical events unfolded in the way that they did. In scientific terms, they are interested in understanding the mechanisms involved in bringing about specific historical events, or, put simply, identifying causes of events.

Jared Diamond, in his book *Guns, Germs, and Steel*[1] – which is basically a book about human history on Planet Earth in the last 13,000 years (nothing more, but nothing less either) – expressed his wish that human history become a scientific field of study, and not, as the common sentiment is today, something unapproachable by the scientific method. While some scientific procedures, such as the radioactive carbon [^{14}C] dating technique, have now been widely accepted by historians, a term used here to include also archeologists, anthropologists, and several other related disciplines, there have been two main arguments against using the scientific method of gathering preliminary data, forming hypotheses about mechanisms, and testing their prediction in historical research. The first is that the subject is just too complex, so that every historical event is *sui generis*, the result of a unique set of a previous constellation of events that constitute the causes of the present event under study, and therefore no general methods can be applied to its study and no general conclusions can be drawn from such study. The second objection is that humans have free will, and therefore human history is not, indeed cannot be, deterministic. "Human choices made this history happen" is a typical comment.[2]

A specific variation of the complexity *cum sui generis* argument is the concept of "Great Men" in history. This concept posits that the personalities of charismatic leaders "cause" certain historical events – that is, the leaders' personalities are the root cause of their behavior. The Great Men concept, which of course includes some great women as well, does not necessarily deny that human personalities are shaped by both their genes and their environment (although people often vehemently disagree on the proportional contribution of nature and nurture). However, while the lives of such Great Men are fertile grounds for biographers, attempts to understand exactly how biological, environmental, and cultural factors contributed to historical events by shaping the personalities of movers and shakers of human affairs, and every other person in society, are generally discouraged by historians, given the multiplicity of such factors and the enormous complexity of their interactions. In essence, historians, both those emphasizing the actions of singular people as well as those inclined to give more weight to the actions of the multitudes, limit their study to proximate causes of historical events – the direct action of people – and refrain from studying how each person at the time that his/her action under study occurred were themselves "caused", or came into being, by previous events.

In short, whether attributing the cause of historical events to a few prominent human individuals or to popular movements, historians generally avoid studying causes of causes at too many steps removed. But by not pursuing root causes, historians are unable to draw general conclusions that would point to the operation of general "natural laws," an outcome that historians often appear to approve of rather than lament. Thus, Thucydides' conclusion[3] that the "real," or ultimate, cause of the Peloponnesian War was not the specific military incidents that preceded it, but instead "What made the war inevitable was the growth of the Athenian power and the fear which this caused in Sparta" is sometimes admired as a stroke of analytical genius but is as often derided as a presumptuous and unbecoming statement from a historian. Overall, the pervasive impression from present-day historical research is that history is not deterministic but is rather a stochastic, random process, and therefore that history cannot be predicted or even retroactively explained by any scientific methodology.

As Diamond points out, there are several well-established scientific disciplines that have a major historical component, such as astronomy, geology, and, of course, biology. What they all have in common is having as their subjects of study complex systems that keep changing their states because of previous actions. Oxygen molecules are the same now as they will be a million years from now and will behave the same way in response to other chemicals (depending on temperature, pressure, etc.) at both times, and therefore a simple equation can predict their behavior anytime. On the other hand, any living human being or another animal, a plant, or even a bacterium is a complex system that changes from one minute to another, because each constantly carries out biochemical reactions that change its chemical composition, internal structures, external shape, temperature, etc., and therefore will not behave the same way over time. A geological formation, such as a mountain, also changes over time (for example, by erosion), so its properties and "behavior" (e.g., how it retains heat or deflects wind) will also change over time.

Nevertheless, leaving human beings aside for the moment, all scientists treat such complex systems, both inanimate and living, as systems that can be studied by the application of the scientific method, so that their behavior in the past can be explained mechanistically and their future behavior can in principle be predicted, given sufficient knowledge. For all these systems, the concept of free will does not apply, and so any observed situation in these areas of study is thought of as the outcome of deterministic processes, the end result of a chain of causes and effects. Such chains of events are indeed often extremely complicated and therefore present conditions of a given system often defy – for now – complete mechanistic explanations, and futures states of such systems are often not amenable to accurate predictions. For example, atmospheric scientists still do not understand well why specific ice ages occurred when they did and how they played out, and they cannot predict the occurrence of the next ice age and its particular length and severity. Nor can geologists predict with any accuracy when and where earthquakes would occur or their strength. However, these shortcomings are attributed simply to lack of sufficient information as well as the inability to conceptually and technically analyze all the data already available. The argument that these issues are simply not amenable to the scientific method is not entertained by any serious scientist.

Although a general scientific definition of "complexity" does not yet exist, a strong argument can be made that living organisms constitute the most complex systems on earth in the sense that they are made up of the largest number of distinct parts – tens of thousands of unique genes, proteins, and metabolites – that interact with each other in multiple ways. Before Darwin published his book *On the Origin of Species*[4] in 1859, in which he promulgated the principle of biological evolution by natural selection, the descriptive study of living organisms – their components and their interactions with each other – was called "natural history." This term is still sometimes used in this way, for example in the designation of natural history museums, while the study of the interactions among living organisms with each other and with the environment has evolved into the scientific discipline called ecology. Once the validity of the process of biological evolution was established, the term natural history came to describe, in addition to the aspects detailed in the original definition, also the study of the origins, evolution, and interrelationships of organisms. By that extended definition, human history is clearly a subfield of both natural history and ecology.

To be sure, for most people the motives for studying ecological systems that do not include people versus studying those that do are fundamentally different. Even ecologists don't really want to write a book of a thousand pages describing how a geographical area with its myriad plants, animals, and microbes changed from year to year from say 500 BC to 500 AD. At best, they want to learn the principles that govern changes in such a system, so we can use this information to predict ecological changes today. But if this geographical area was the British Isles, and we were historians, we would likely try to record all "major" events, and some minor ones, that occurred there and involved humans during this time. We will note some changes to the system as a whole – for example, deforestation and the gradual elimination of many non-domesticated animal species – but only because humans were the causative agent (or so we imagine), and because such changes had in turn an effect on human affairs later on. "Random" natural events such as earthquakes, floods, or droughts will also be noted, but again only for their effects on humans, and would therefore be labeled as "natural disasters." Basically, we will define our interest as "human history," not ecology, and the interest of our readers would most likely be based simply on the need to relate to their own ancestors (if they are British) or to generally satisfy their curiosity about the past of our conspecifics, not to mention political needs such as establishing national standing and territorial claims.

A "scientific" understanding of human history – abstracting the rules from specific cases – is not the major driving force in conducting historical research and writing it up, nor in reading historical research. But as I argue in this book, a scientific understanding of human history is not possible when the roles of all other components in the ecological system are minimized or completely ignored, and humans are considered as the sole causative agents of any action.

Besides the claimed insurmountability of the complexity problem, the second objection to making human history part of the scientific discipline of natural history, as mentioned earlier, is the idea of free will, which sometimes goes together with the blanket denial of any biological basis for human behavior. As noted above, humans as well as other living organisms are complex systems that are constantly changing. The cells that make up the human body grow and divide, and sometimes die, in response to both genetic and environmental cues. Information from the body itself and from outside the body is obtained by the nervous system and is stored in the brain, which undergoes physiological and chemical changes in response to these inputs. Given the complexity of the system, it is indeed difficult to predict what output a given input will elicit. Fair enough, but the concept of free will claims something fundamentally different. The concept of free will posits that output (human action) is independent of input – or, to put it another way, given a single defined input, multiple output choices are possible. If correct, it means that a person can somehow "choose" his action unconstrained by the condition of his brain, whose chemical and physiological state at the time the choice is made had been determined by the interactions of all past inputs, up to and including the said input, with the "hardware." This concept is obviously at odds with the basic physical laws of nature, denying as it does that there is a physical cause, and only a physical cause, for each action, and has therefore been refuted by scientists beginning with the ancient Greeks. Nevertheless, the belief in free will is pervasive among laymen and even scientists.[5]

With the concept of free will disposed of, there is really no reason why in principle human history should not become a sub-discipline in both scientific fields of ecology and natural history. Perhaps this approach has been most closely adapted by those anthropologists who call themselves sociobiologists. Sociobiology, as defined by E.O. Wilson,[6] is the study of social living organisms based on population genetics and evolutionary biology principles. Sociobiologists note that even learned behavior requires some hardware – some biological structures such as a brain – for it to be acquired and executed, so any genetic variation that leads to differences in that hardware could cause differences in behavior among individuals, and therefore could be subject to natural selection. The sociobiological approach does not specifically address how much "nature" versus "nurture" there is in any specific behavior, except to reject the extreme notion, still present in some cultural anthropology circles today, that all human behavior is learned and is transmitted among people absolutely independently of human biology, an untenable position since any human behavior, as noted above, has to reside in a biological system. The argument that cellular activities underpin behavior should not be turned into caricature statements such as "there is a gene for liking classical music" or "there is a gene for being a murderer." Most behaviors are the results of the interplay between multiple genes as well as many environmental inputs.

Since the argument that "I am aware, or conscious, of making a decision, therefore I have free will to choose" is both scientifically untenable and inconsistent with experimental evidence, it follows that explanations of historical events that invoke "human choice" as ultimate causes cannot provide a complete, mechanistic explanation to human history. The analysis of the historical events described in this book will therefore be based strictly on materialistic arguments for cause and effect that are not dependent on human awareness. Such arguments, however, should not be confused with dialectic materialism, based on the writing of Marx and Engels (neither of them used the actual term, although Marx wrote about the "materialistic conception of history"). These two intellectuals knew little biology and in particular nothing about the biological basis of human behavior, and therefore, while correctly describing many "problematic" social conditions (problematic in the sense that at least some people consider these conditions to violate certain moral and social codes), often misidentified the ultimate causes of such conditions and proposed (or predicted) unrealistic courses of action to change these conditions. In keeping with the scientific treatment of history in this book, it is important to note that this book does not define anything as a "problem" and therefore does not propose any solutions. Instead, I attempt here to objectively present facts, identify patterns, and draw general conclusions about how the natural world operates under such circumstances. It must be emphatically stressed that such an approach should not tempt the reader to draw any conclusions concerning my own notions about the "justness" of the identified mode of operation of the natural world which humans are part of, and certainly not see it as a moral endorsement. Far from it – particularly, but not exclusively, when violence is involved. But morality is not an issue within the purview of science, and of this book.

From the ongoing discussion above it is fair to conclude that human behavior is a subject that falls within the scope of ecology, as does the behavior of any other living organism.[7] In this book I will therefore definitely adopt the sociobiological framework. Contrary to the often-expressed sentiment that we should avoid turning to nature to explain human events, my argument is that there is no other way to fully explain human behavior (and culture generally) except by examining nature. As pointed out by Richard Dawkins in his influential book *The Selfish Gene*,[8] the evidence is clear that humans behave in principle like any other living organisms, driven by selfish genes and subjected to natural selection. However, by "examining nature" I do not mean only human nature, but the entire natural milieu that humans are embedded in. In fact, as my training is neither in human biology nor in history but in plant biology, the main focus of this book is to describe and illuminate specific aspects of plants that in my opinion have played very significant, indeed inordinate, roles in influencing the course of "human" history. Many of these aspects, of course, have already been studied, analyzed, and extensively written about, most notably the transition of human societies to farming, which depended largely on the domestication of plants and, to a lesser extent, on the domestication of animals. It is fair to say, however, that general history books for the lay public, even when containing a fair amount of description of agricultural issues or other events in which plants played an important role, were not written by plant biologists. Such books understandably are written with a focus on people, and with a minimum of specific information about the plants themselves. In this book, I hope to bring plants to the fore, and show that their contributions to history have been substantial.

However, even a plant biologist like myself would find it difficult to write a natural history of the world with plants, not humans, at the center. The conviction that humans occupy the center of the biological world appears to come naturally to members of the *Homo sapiens* species, and clearly it is a conviction that is very difficult to avoid embracing. So my goal is a bit more modest than attempting to offer a completely new way of looking at history. What I hope to achieve here is to provide the readers with more detailed and nuanced descriptions of the involvement of plants in historical events. Such a description will necessitate presenting a great deal of information about the biology of plants that is usually not found in history books, with the aim of showing that specific aspects of plant biology, often not fully known to the general public or professional historians, have played crucial roles in human history. In this retelling of history, plants will be, if not at dead center of the story, at least a bit closer to the center than usual. Furthermore, I do not intend to be comprehensive. Instead, specific historical incidents have been chosen that represent good opportunities to demonstrate the important contribution of plants to human history. Such chosen examples already have a wealth of historical information available, and the added information here about the role of plants in these events is meant to enrich the analysis and to offer, when possible, new angles for causal explanations of the observed chain of events. Secondly, I have chosen to further restrict my examination mostly to the role of plants in violent, lethal human conflicts, or wars, because wars are a type of human behavior in which biological principles are most starkly at play. Indeed, it is ironic that plants, which unlike animals are considered incapable of violent activity, have been so heavily involved, both as causes as well as means, in human strife.

WAR AND LIVING ORGANISMS

Darwin was not the first one, of course, to point out that life, for any individual of any species, is a struggle. However, his overarching scientific theory of how natural selection has shaped the interactions between living organisms and their long-term evolution, published in 1859, provided a theoretical foundation for countless scientific experiments and observations that have now amply confirmed the validity of his theory. Darwin did not know about the work of his contemporary Gregor Mendel, who experimentally discovered the basic units of hereditary that were only later named "genes" by a Danish biologist, Wilhelm Johannsen, and it was not until the middle of the 20th century that genes were shown to be comprised of the DNA double helix molecule. Nor did he know the details of sexual reproduction. But he correctly surmised that life forms with different combinations of genes evolving over time as the result of genetic changes occurring by chance in different individuals, combined with differences in the relative success of genetically different individuals in producing progeny – passing one's genetic traits to the next generation. This relative rate of reproductive success is defined as biological fitness, and it is determined by a process Darwin called "natural selection." For unicellular organisms that reproduce simply by cell division, fitness depends on the ability of the individual to obtain from the environment resources for survival and growth. But for organisms that engage in sexual reproduction (see Glossary) – the dominant form of reproduction in most animal species and to a lesser extent in plant species – fitness also depends on the ability to obtain access to mates.

With our extensive knowledge about genetics today, we understand that individuals are genetically different from each other and from their parents for two reasons. The first is that genes sometimes slightly mutate, or change, and thus work slightly differently from the original version of the gene (gene variants are called "alleles"). Second, in sexual reproduction, each individual progeny inherits one complete set of genes from one parent and a second complete set of genes from the second parent. While each parent has two sets of genes (i.e., two copies, or alleles, of each gene, with the two copies being either identical, or not), the single set of genes a parent contributes to the progeny is assembled by taking half the time the gene copy from its set #1 and the other half of the time the gene version from its set #2. Since the particular assembly of the gene set that each parent contributes is different in the formation of each progeny (basically, which version of each gene is included in the genetic endowment that the parent provides each progeny is determined by luck), sexual reproduction thus gives rise to progeny each with a unique gene combination that is not identical even to its siblings.

Because for every given species at any given time, genetic variations among individuals exist and some combinations render individuals more successful at reproduction than others, in the aggregate, the new generation will comprise of individuals whose genetic makeup is not the same, again in the aggregate, as that of the older generation. Instead, the genetic material of the new generation is derived disproportionally from a subset of individuals of the parent generation.

Furthermore, the struggle for resources and mates is unending. It continues in every generation and is always engaged by individuals with different genetic resources, so there will always be winners and losers. However, it is important to realize that reproductive success, or fitness, is not a fixed trait because environments (i.e., anything that is not the actual individual itself, such as other living organisms and inanimate objects that the individual interacts with) change, and therefore different genetic resources might work better in different environments. And once groups of individuals (called "populations," see Glossary) from the same species find themselves in different environments that favor different gene combinations, the genetic changes in such lineages over time can be so extensive that different populations become distinct species whose members cannot interbreed if and when their living areas overlap again.

Individuals of the human species, like all other living things, compete for natural resources with other individuals of the same species and with other species. In addition, since we are a social species, we often band together to form groups that compete with other human groups as well as with individuals and groups of other species. The use of brute physical force to obtain resources is a basic approach for all animals in their interactions with other species, with deception, including mimicry (see the example with orchids later in this chapter) being a distant second. The use of force to obtain resources is also common among animals in their interactions with conspecifics, and this observation includes social animals such as humans, although with social animals there are more opportunities for deception. Even within organized, peaceful human societies where intragroup competition appears on the surface to be entirely peaceful and conducted according to established rules, this is only so because individuals know that when the rules are violated they will be enforced by physical force. As for intergroup interactions, throughout history wars have been so common between different human populations that it can be argued that this is the "normal" state of affairs. And regardless of the proximate causes provided by the belligerent groups, the origin of violence and wars can always be traced to competition over one type of resource or another, and ultimately to the basic need of humans to obtain mates and to ingest organic material.

The anthropologist Napoleon Chagnon has proposed, based on his work with the Yanomamö tribe in the Amazon, that human wars began as a way for males to acquire females of reproductive age.[9] He observed that groups of Yanomamö men often attacked neighboring villages and abducted their women. Moreover, Yanomamö men who were fiercer, as proven by the fact that they had killed other people, had more wives and sired more progeny than men who had not killed. Similarly, the wars of the nomad Mongols, who about 1,000 years ago reached all the way from China in the East to Europe in the West, were basically raids for women and loot.[10] The Mongols were not, at least initially, interested in acquiring territory. Instead, when they conquered a city they would kill all the men, rape the women, and haul all the booty back to Mongolia. When their leader, Genghis Khan, was not on site to participate in the raping himself, his forces would make a collection of choice women and send them to him. A recent scientific study[11] found that 8% of men living in the area once under the Mongols' rule (or 0.5% of the total male population in the world) are direct descendants of Genghis Khan (men's origin was examined because it was easy to follow genetically, but the same percentage of women is expected to belong to Genghis' lineage). Throughout history, and even today, rapes by soldiers during wars around the world have been common and have often served as an implied incentive and reward for service (as was loot).[12]

Human wars have always been carried out almost exclusively by males. This is consistent with the origin of the warring behavior in humans as a form of competition among males for females – a man can have more children if he has more mates, but this is not true for women. But males have continued to maintain almost complete exclusivity in participation in wars,[13] even after mate recruitment evolved into more peaceful procedures in societies that have larger population size and therefore do not rely on raids of other groups to obtain women.[14] The reasons for this may be complex and are not within the scope of this book. However, for our purpose we will note that all large-scale wars between parties with defined borders have been about the second aspect of the competition of life – to obtain material resources to allow people and their progeny to survive and thrive. It is true that specific wars were often claimed to be undertaken for "revenge," "national honor," or religious reasons. However, regardless of the beliefs and public declarations of leaders who wage such wars and the soldiers who follow their orders as to their reasons for fighting, on closer inspection wars are initiated by the aggressive party to establish dominant power over another group so that it can enhance or maintain its ability to obtain resources to the detriment of the vanquished party.

While plants and animals are both living organisms, they clearly have different body plans and overall look quite different. Plants also, for the most part, do not move. But these superficial differences mask the vast similarity in the biology of plants and animals. Both animals and plants are made of cells. Both also have distinct sets of genes (called "genomes") which provide the "program" that determines, via interaction with the environment, how their lives proceed from their start as embryos to their death. The genes determine how cells divide, organs are formed, how the body grows and develops, and how the body is kept alive. And many of the genes found in plants are basically the same as those found in animals, as proven by the fact that scientists have successfully replaced defective animal genes with the corresponding genes from plants, and vice versa.

Plants also reproduce sexually like animals.[15] They have male and female organs, and just like animals they produce sperm and egg cells, respectively, each containing one set of genes contributed by the parent, as described above. In both plants and animals, the fusion of an egg cell with a sperm cell leads to the development of an embryo that will eventually grow into a mature individual. However, in plants reproduction is more versatile than in animals. The largest group of plants, called angiosperms, or flowering plants, may also engage in a variety of both sexual and non-sexual reproductive methods beyond the basic sexual reproduction process described above that involves a male individual and a female individual. Many angiospermous plants are hermaphroditic, meaning that they have flowers with both male and female parts (there are a few hermaphroditic animals too). The male parts in plants – the stamens found at the end of the filaments of the flowers – produce pollen grains, each carrying a male sperm cell, or "gamete," inside it. The female parts of the flowers include an ovary where the eggs (i.e., female gametes) are present, and a style extending from the ovary and culminating in a stigma. Once pollen grains leave the anther and land on the stigma – a process called pollination, which can be accomplished by wind or by an animal pollinator – each pollen grain germinates and grows a long pollen tube through the stigma and style to deliver its sperm to the egg, and the fusion of egg and sperm (a processed called fertilization) begins the development of a new plant embryo. In flowering plants, the plant embryo is encased by tissues that develop from the ovary, creating a seed. Embryos often dry out, enter dormancy, and stop growing, but once the seed enters the soil and imbibes water the embryo begins growing again and develops into a mature plant.

Plants with perfect flowers (Figure 1.1), defined as having both male and female parts, can sometimes fertilize themselves, but in other cases they are "self-incompatible", meaning that the sperm-bearing pollen must come from another plant because the plant has evolved a mechanism to prevent its own pollen grain from growing through the stigma and style and delivering the sperm cell it harbors to the egg. Other flowering plant species may have separate male and female flowers on the same plant, or separate plants with only male or only female flowers. Non-flowering plants, such as ferns, have evolved somewhat different mechanisms for sexual reproduction, which will not be described here.

Most importantly, though, a large number of plant species can reproduce nonsexually. For example, plants often grow shoots out of their roots, and when the roots bearing the new shoot are cut off from the original plant, an independent plant is

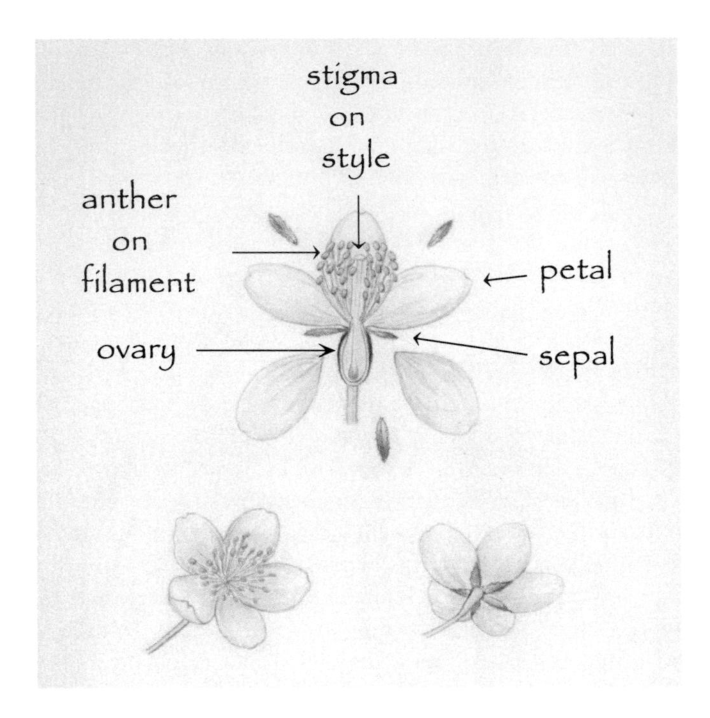

FIGURE 1.1 The flower of the cherry tree is a perfect flower. The pollen grains, each containing a male gamete, are present in the anther, which is a structure present on the end of the filament (the combined structure is called a stamen). The female gametes, or eggs, are present in the ovary, from which extends a style that terminates in a stigma. The entire complex of female structures – stigma, style, and ovary – is called a pistil. Modified leaves called sepals and petals flank the male and female structures.

created that is genetically identical to the one from which it originated. Additionally, a shoot that is cut off from a plant and buried in the ground may start growing roots and become an independent plant, also genetically identical to the plant from which it was derived. Such non-sexual propagation of plants, which creates genetically identical progeny, or clones, is often referred to as vegetative reproduction.

To grow in size and to replace old and defective material in its body, a living organism must continuously acquire carbon-containing molecules, termed organic compounds, that are used both as building material and as fuel to generate the energy for the chemical processes involved in body growth and maintenance. But while the biochemical processes of building and maintaining the body are quite similar in animals and plants, the sources of the building material and the initial generation of the energy to make these building blocks are distinctly different. Plants harvest energy from the sun in the process of photosynthesis, and they use this energy to convert carbon dioxide gas from the air and water from the soil into organic chemical compounds. These compounds are then used to carry out metabolism just like animals do, employing enzymes that the cells make to speed up the chemical reactions. Such enzymes often need minerals such as iron, zinc, or calcium to work properly. Therefore, all that plants need for growth is water, minerals, carbon dioxide, and sunlight. Animals, on the other hand, cannot photosynthesize and therefore must acquire organic material by directly eating plants or by eating other animals that eat plants or that eat other animals that eat plants etc. Ultimately, all animals obtain key organic compounds – those that they cannot make themselves and that serve as the building blocks for other compounds – and most of the minerals they need from the primary producers, plants (or photosynthetic microorganisms and algae). Moreover, oxygen is a byproduct of photosynthesis, so plants not only provide food for animals, but also oxygen for breathing. Plants are therefore said to be at the bottom of the "food pyramid." On the bottom, yes, but because of the process of photosynthesis, plants are a uniquely productive class in our global society of living organisms.

Actually, plants are not as passive as they seem. We tend to think about plant domestication as a process that humans engage in to help us obtain large, reliable, and steady sources of food. The surplus food produced through the development of agriculture no doubt helped humans to increase in numbers and achieve population densities that led to the rise of urban centers and specialized professions engaged in activities beyond the simple production of food. But does this mean that plants were the losing party in this process? Consider this: Every year, there are by far more rice plants growing around the world – with a much larger geographical range than that achieved by wild rice – than there are people. And rice as a species is doing so well because humans obtain and prepare land for its growth, sow its seeds in the ground, fertilize the growing plants, protect the plants from other plant competitors (usually by weeding) and from animal pests (usually by spraying pesticides), and collect the seeds once they are produced and save and protect enough of them to plant next year, all at considerable expense of energy to the humans involved in this enterprise.[16]

Humans in fact are not the only animal species that plants have domesticated. While rice pollen is transferred from one plant to another by wind, many plants need animals to transfer their pollen from one plant to another. Plants have evolved elaborate mechanisms to get animals to pollinate them, for example, producing scented and colored flowers to attract insects, birds, and some mammals (such as bats) to the flowers. They also often provide small amounts of chemicals, such as sugars and amino acids in the flowers, as rewards for visiting pollinators. And sometimes, they simply fool male insects into visiting the flowers and carrying pollen from one flower to the next by looking and smelling like female insects "in heat" – a strategy that has been termed "pollination by sexual swindle" and in which orchid species seem to be particularly good at.[17] As for our "domesticated" crops, while humans do not fly from one flower to another to pollinate them, farmers do regularly contract bee keepers to place beehives next to fields of crop plants that need insect pollinators. And in fact hand pollination, often with a brush, is a common agricultural practice for obtaining seeds and fruits from valuable plants that have been introduced by humans into a new region outside the range of the plants' natural pollinators.

The co-option of animals by plants is not limited to the pollination step at the beginning of the reproductive process. As we will see in later chapters, a plant seed contains both the embryo as well as some food reserves, and sometimes the food reserves are quite large and in a such a case they are called a "fruit" (for now, we will stick with the lay term for fruit; the botanical term will be defined in Chapter 2). Animals often ingest fruits with seeds or just the seeds themselves, and at least a portion of the seeds survive their passage through the animal's gut and are deposited by the animal away from the position of the mother plant, thereby helping the plant disperse its progeny. Sometimes, the animal will spit the seeds rather than swallow them – another way to disperse them. Some animals actually engage in planting the seeds in the ground, such as squirrels and some birds that bury seeds and nuts in the ground to create stashes of food for later use. When the animal fails to retrieve them on time, the seeds often germinate.

Humans also engage in large-scale vegetative reproduction and dispersal of plants – as we will see in later chapters, sugarcane and banana plants, for example, are propagated by cuttings. More recently, people have developed a "micropropagation" process in which millions of new plants are generated from a single plant cell. Overall, plants have been extremely successful in getting animals, and particularly humans, to propagate them, disperse them, cultivate them over larges areas of the globe, and ensure that they thrive. In facts, people will kill to make sure that their plants are successful, as we will see throughout this book.

As someone who grew up in Israel and served in its military, I am quite aware that an armed conflict over a territory is a classic example of a fight over resources between two human groups. People need territory to grow plants or animals (that eat plants) for food. Water resources are also important for agriculture, so it is not surprising that wars over water have occurred throughout human history. Wars over other resources – territory that has crude oil, for example, or metal ores or other minerals – might appear at first to be more about standards of living rather than subsistence, but a closer inspection would show that this is not so. Since all life ultimately depends on growing plants, the first purpose of obtaining oil, coal, and iron, to name just a few resources, is to maintain a society that can continue to carry out food production so it can continue to propagate itself into the future. For examples, iron is essential for making agricultural implements (and weapons too), and energy is essential for making fertilizers and for food processing and preservation in general. A society that stopped reproducing would be a society that no longer fights wars.

So plants are the foundation of our existence and the ultimate cause of our wars. Fighting over possession of land in which specific plant species, such as those used to make spices, grow has also been common throughout history, as well as fighting to prevent one's opponent from obtaining such resources. But plants are not only the cause of our violent conflicts, or wars – they also provide us with means to carry out such conflicts. After years of studying plants and the valuable compounds that they make, I have become aware that many different plants have played important and direct roles in numerous human conflicts. Rubber, a product of the tree *Havea brasiliensis*, is a crucial component in modern machinery, including weapons. Drugs against malaria, which are obtained from certain plants, have enabled armies to invade areas where they would not have survived otherwise (for example, the US marines in the Philippines). Many war implements, such as bows and arrows, were until very recently constructed from wood. Naval warfare would have been impossible, at least until the end of the 19th century, without the trees used to construct the ships and the plant pitch used to make these ships seaworthy. And coal, produced by geological and chemical processes from dead plants, and oil, similarly produced from mostly dead marine photosynthetic organisms, have been crucial energy sources for the construction and operation of the machinery of war in the last few hundred years.

Perhaps some words on the use of the term "war" are in order. Humans often carry out aggressive campaigns against plants and animals, but only when operations also involve human opponents are these campaigns called "wars." Farming itself is a human practice in which some plant species are exterminated and others are cultivated and allowed to grow. It is not generally considered a war, but farmers will often talk about their war against weeds and animal pests. On the other hand, the extermination of opium poppy fields in Afghanistan or coca plants in Colombia has certainly been considered as part of the "war on drugs," and such operations share substantial characteristics with conventional modern wars and are often carried out by military personnel using military equipment. Even more directly related to war, plants, plant products, and food derived from plants are often attacked and destroyed to prevent the enemy from being able to use these resources, or, alternatively, appropriated for the benefits of one's own army and population.

Generally, history books do not dwell on the specific characteristics of the plant species involved in wars. Plants are described in general terms as food sources, building material, etc. The emphasis is of course on people, their actions, their needs, and the consequences to humans. However, my long acquaintance with plants through my own studies and my reading of studies by others has convinced me that a better knowledge of the traits possessed by the different plant species involved in human wars will actually contribute toward a better understanding of causes and effects in human history. Accordingly, this book will tackle such issues as what properties of spice plants made them so valuable that in the not-so-ancient past several European countries got involved in major wars over possessing the lands in which they grow, and how sugarcane cultivation contributed to the establishment of slavery in the New World. On a more fundamental level, the material covered in the following chapters will show how the establishment of agriculture made human warfare virtually unavoidable. In summary, I hope to show that plants have played more important roles in historical events than we typically appreciate, and that a more detailed understanding of plant biology will lead us to a much richer understanding of events in human history.

NOTES

1. Diamond, J. 1997. *Guns, Germs, and Steel.* W.W. Norton & Company.
2. This quote is from Hahn, B. 2011. *Making Tobacco Bright.* John Hopkins University Press.
3. Thucydides, *History of the Peloponnesian War*, translated by Rex Warner. 1954, Penguin Books.
4. Darwin, C. 1859. *On the Origin of Species by Means of Natural Selection, or the Preservation of Favoured Races in the Struggle for Life.*
5. See Cashmore, A. R. 2010. The Lucretian swerve: The biological basis of human behavior and the criminal justice system. *Proceedings of the National Academy of Science USA* 107: 4499–4504. While the concept of "free will" appears to be something that people instinctively believe in, since they have the feeling (i.e., consciousness) that they are making decisions, free will is an explicit tenet of most religions. It solves the problem of holding people responsible for their action in the face of the dogma that God is all-knowing and almighty, and therefore the future is preordained since nothing can

happen against God's will. Religions therefore promulgated the inconsistent concept of free will to counter the excuse that "God (or the devil) made me do it." Secular societies have retained the concept in their moral and legal systems because people naturally feel that it would not be fair to punish people for acts that they committed if they had no choice but to commit them. This is a long and complicated issue (adverse consequences brought about by societal action to people who commit acts that go against the interest of society at large could be justified by various other ways, even when the perpetrators had no control over their actions; see discussion in Cashmore, 2010), and one that both educated and uneducated people try very hard to avoid. Darwin, who recognized that free will was a fiction, remarked that "this view will not do harm, because no one can be *fully* convinced of its truth, except man who has thought very much." In other words, as long as this knowledge is confined to a few intellectuals, human society is safe.

6. Wilson, E. O. 1975. *Sociobiology: The New Synthesis.* Harvard University Press.
7. All living organisms display "behavior." Scent emission from flowers is no less a behavior than an insect flying toward a scent-emitting flower. Ecology is the scientific field that studies the outward manifestation of the behavior of living organisms, i.e., the interactions among living things as well as their interactions with inanimate objects, and their consequences. The study of the internal mechanisms that generate behavior, however, is different for different types of organisms. For plants, this area usually falls under the disciplines of plant physiology, plant biochemistry, cell biology, and similar fields that are defined more for historical reasons than for real presently valid intellectual reasons. When it comes to animal behavior, and particularly human behavior, the disciplines that study its internal generation are more numerous, are sometimes overlapping, and often compete for legitimacy.
8. Dawkins, R. 1976. *The Selfish Gene.* Oxford University Press.
9. Chagnon, N. A. 2013. *Noble Savages: My Life Among Two Dangerous Tribes – The Yanomamö and the Anthropologists.* Simon & Schuster.
10. De Hartog, L. 1999. *Genghis Khan, Conqueror of the World.* I. B. Tauris Publishers.
11. Zerjal T. et al. 2003. The genetic legacy of the Mongols. *American Journal of Human Genetics* 72: 717–721. But this does not mean that Genghis Khan and his soldiers, or the Yanomamö for that matter, understood Darwin's "evolution by natural selection" theory or the concept of fitness, and were consciously and deliberately acting to increase their fitness. Quite the contrary – as I argued above, awareness, deliberation, and "making choices," even if they appear to exist, are not the causative agents of human behavior. And there are always exceptions to the rule. Alexander the Great, who embarked on his quest to conquer the Persian Empire and the rest of world with his male lover Hephaestion, was known for treating high-born enemy women with respect. While he nonetheless acquired several such women as his consorts during the campaign – presumably by mutual consent, or at least the consent of the women's families – he totally failed (ignoring kin selection here) as far as evolution is concerned, as his only verified progeny, a son by Roxana, died without issue.
12. Keegan, J. 1993. *A History of Warfare.* Vintage Books.
13. It remains to be seen if recent changes in policy among militaries of Western countries that allow for the participation of women in combat will hold in the long term.
14. All human groups in general avoid matings between closely related individuals. Such matings often result in progeny with congenital defects.
15. For general background on plant biology, see Raven, P. H., Evert, R. F., and Eichorn, S. E. 1999. *Biology of Plants*, 6th ed. Freeman and Company.
16. A similar view on plant domestication by humans is presented in Pollan, M. 2002. *The Botany of Desire.* Random House.
17. Bohman, B. et al. 2016. Pollination by sexual deception – it takes chemistry to work. *Current Opinion in Plant Biology* 32: 37–46.

2 Fighting Grains

GRAINS ARE TARGETS OF WAR

Human domestication of plants and animals (or vice versa) occurred independently in several places around the globe, starting at least 10,000 years ago. Agriculture – the system of producing food by growing domesticated plants and animals – was clearly a more successful method of obtaining a large and constant supply of calories compared with hunting and gathering. As described in great detail by Jared Diamond in his book *Guns, Germs, and Steel,*[1] societies that adopted agriculture often experienced huge population growth. They were therefore able to successfully out-compete hunter–gatherer societies and take over their territories, either by force or simply by interbreeding with members of hunter–gatherer societies and bringing their agricultural practices with them. As a consequence, the vast majority of the people living today belong to societies that obtain most of their calories from agriculture.

The most obvious need of agricultural societies is land on which to grow their crops as well as their animals, which themselves either graze on naturally growing vegetation or are fed plants grown by the farmers. As agricultural societies began to produce more food than was required for the subsistence of the farmers themselves, professional specialization in non-farming (and farming) tasks also began to develop, urban centers arose, and social stratification evolved to give rise to different castes and to centralized government. Yet, as agricultural societies prospered and out-competed hunter–gatherer societies, each agricultural society found itself competing for land resources with other agricultural societies. Indeed, fighting over land has been the most common reason for war throughout history, and thus an additional social development occurred – the creation of a professional army. Because wars over (agricultural) territory have been so common, such wars will not be discussed here per se. Instead, we will examine the plants that are the major contributors to the sustenance of human and domesticated animal life, explore the properties of these plants, and consider how these properties have made these plants both the cause and the target of warfare.

It is worth restating the obvious fact that humans, like all living organisms, are a type of machine that obey all natural laws, foremost among them the second law of thermodynamics. In lay terms, the implication of this law to humans is that we must continue to ingest combustible material (i.e., food) so that our internal metabolic processes can continue to produce the energy to keep our bodies alive as well as to produce movement through our muscles.[2] In wars throughout history, each side often attempted to cut off supplies of food to the enemy or to destroy enemy food supplies, with the enemy often being defined as the entire population and not only combatants. This was done for the obvious reason that once the enemy is deprived of food, their physical condition will worsen and they will not be able to continue fighting well and therefore be defeated easily. Even if the starved enemy did not lose a battle or surrender but instead, hid inside walled cities, it was just a question of time until the besieged defenders starved to death and the surrounding army won the war. Typically, serious health effects occur within 35–40 days of complete cessation of food intake (given that water intake continues normally), and death occurs within 40–60 days, after various essential metabolites in the body are completely exhausted.[3]

The destruction of foodstuffs as part of war has inflicted such a huge toll on human life and caused widespread misery throughout history, with civilians being the majority of the victims (when food is short, the people without the weapons are the first ones to experience starvation), that this military tactic has now been officially banned under Article 54 of Protocol I of the 1977 Geneva Conventions. This article states in part:

> It is prohibited to attack, destroy, remove, or render useless objects indispensable to the survival of the civilian population, such as foodstuffs, agricultural areas for the production of foodstuffs, crops, livestock, drinking water installations and supplies, and irrigation works, for the specific purpose of denying them for their sustenance value to the civilian population or to the adverse party, whatever the motive, whether in order to starve out civilians, to cause them to move away, or for any other motive.[4]

However, the Geneva Conventions and this particular article are very recent developments in human affairs, and it remains to be seen if such international treaties will actually have a measurable effect on the conduct of future wars. Indeed, the treaty is binding only for those countries that sign and ratify it, and many countries, including the United States (which is currently involved in more armed conflicts than any other country in the world), have not yet ratified this Protocol. Whatever the future holds, a quick perusal of history books should be sufficient to convince us that the tactic of attempting to deprive one's enemy of food (and often other resources as well) has been extensively used, often to good effect. There are numerous examples of armies and cities that came under siege and had to surrender for lack of food.

An ancient, complex situation for which we have detailed historical information is the Peloponnesian War, which occurred in 431–404 BC between Athens and her allies and Sparta and her allies. Athens was a naval empire, and while its original territory of Attica, where the city of Athene was (and still is) located, was on the mainland, its citizens had direct ownership of land in many Aegean Islands. In addition, many Aegean Islands, as well as Greek cities along the Asia Minor coast and the Black Sea coast, were voluntary or involuntary members of the Athenian Empire.[5]

When war broke out between Athens and Sparta, Spartan forces invaded Attica at the beginning of each campaign season,[6] devastating the countryside and ruining all crops. This, however, had little effect on Athens, since the city was protected by walls, the majority of the rural population withdrew from the land and moved into the city, and Athens's port at Piraeus was still functioning and able to receive grain shipments from Athens's colonies. Indeed, even before the war began Athens's population was of such a large size that the local food production in Attica, the region where Athens is located, was not sufficient to accommodate its needs. Food had to be brought in from the Athenian citizens' plantations in the Aegean Islands and from lands cultivated by others further east as far as Crimea. The main type of imported food was wheat, and it was these grain shipments that the Spartans tried to stop with their own warships during the Peloponnesian War. While Spartan admirals were occasionally successful in intercepting grain ships going to Piraeus, for much of the war the Spartan navy was no match for the Athenian navy, and Athens did not suffer serious food shortages. However, this situation changed in 405 BC, after the Spartans appointed a new and talented admiral by the name of Lysander. The Spartan fleet at the command of Lysander decisively defeated the Athenian fleet at the Battle of Aegospotami in the Hellespont (present-day Dardanelles in Turkey), the straits that connect the Aegean Sea to the Black Sea. Once defeated, unable to secure its grain shipments, and facing certain starvation, Athens surrendered.

Until recently, destruction of crop plants in the field was for the most part a laborious and time-consuming work given the inefficient tools that people had developed to carry out such operations. Plants needed to be uprooted, trees had to be cut down. However, if the plants could be burned, that was another story. But to be fit for burning, the plants had to be sufficiently dry. This was a problem with most food crops that are ripe for harvesting while still green. However, fields of grain-bearing plants such as wheat, which die and dry in the field as they mature before the grains (i.e., the seeds) can be harvested, were an arsonist's dream. Not surprisingly, descriptions of the burning of grain fields and grain stocks during military campaigns are numerous.[7] A notable example occurred during the American Civil War. In 1864–1865, General Sherman and his troops marched in several columns from Atlanta to Savannah, Georgia, and later from Savannah through the Carolinas, living off the land and destroying everything they did not need, including killing farm animals and burning stored food stocks as well as crops in the fields. The intended purpose of the destruction was not just to deprive the enemy of these resources but also to terrorize the civilian population and demonstrate to them that their own army could not defend them.

It must be noted that the difficulties with actually carrying out large-scale destruction of crops in the field have largely been solved by modern technology that combines mechanical and chemical innovations. The British were the first to employ such technology in the Malay Emergency[8] of 1948–1960, aerially spraying crops in forest clearings with the herbicide Agent Orange to prevent rebel forces from growing food, mostly sweet potatoes and corn. The British also used the same herbicide to spray over large swaths of jungle to cause defoliation and thus prevent the guerrilla forces with whom they were fighting from being able to hide there. It was this precedent of herbicide use during war that allowed the United States to claim such use legal and to consequently apply the same herbicide on a huge scale during the Vietnam War (involving Cambodia and Laos as well as) in the 1960s and 1970s.[9]

While it is common for an invading army to try to prevent besieged defenders from obtaining food, deliberate destruction of crops and food by a retreating army, a tactic that came to be called "scorched earth," has also been frequently employed throughout history. Invading armies need food for the duration of the campaign, and until very recent time troops carried only limited amounts of food with them and obtained most of their food locally as they advanced, a strategy called "living off the land."

These supply requisitions were done either simply by force or by negotiations with the local population. This method of supplying the army depended on the availability of local food, and this was another reason – in addition to clement weather – why military campaigns were typically conducted during the growing season.

A typical example of the employment of the "scorched earth" defensive tactic from ancient history comes from the campaign of Roman Emperor Julian against the Persians in 363 AD. As the Roman troops crossed the Tigris river and advanced eastward through a generally very fertile land, the Persians drove away their animals and burned their own fields, so that "the grass and ripe corn were consumed with fire; and, as soon as the flames had subsided which interrupted the march of Julian, he beheld the melancholy face of a smoking and naked desert."[10] The result was that the Roman troops could not sustain themselves and had to withdraw.

Perhaps the most famous example of the successful use of the scorched earth tactic in defeating an invading army comes from the invasion of Russia by the French Grande Armée under Napoleon.[11] The invasion began during mid-summer of 1812. The Russian army under Tsar Alexander I chose to retreat rather than engage the French forces, but not before destroying all farm buildings and animals and burning down the fields and any stored food stocks that they could not take with them or bury. The lack of food was a major hindrance to the French army. The soldiers, suffering from hunger, frequently wandered off into the countryside in search of food for themselves and their horses, and were often killed or captured by locals. Nevertheless, the French army reached all the way to Moscow in mid-September 1812, after barely defeating the Russian army in the major Battle of Borodino on September 7, a battle in which the French army also suffered many casualties.

On entering the just-deserted city of Moscow, the greatly diminished French army found plenty of food in the cellars of the abandoned houses, but no sooner had they settled in than numerous fires broke out in the city, began by the few remaining Russians on the order of the Russian military. Although the French troops were still able to hold onto enough food to sustain them for a few months, it was not enough for a stay in Moscow for the entire winter, and therefore on October 18 Napoleon and his troops began a retreat back to Central Europe. The food found in Moscow was packed and transported with the retreating soldiers, but a major difficulty was the lack of food for the horses pulling the supply carts. It was too late in the season for the horses to find grass to graze on, and the army was not able to obtain grain and fodder to feed the horses. The horses began to starve and were not able to carry the loads, and the inevitable consequence was the abandonment of supplies on route and the starvation of the soldiers. At the end, the French army suffered major casualties from the synergistic effects of lack of food and cold weather. The failure of the Russian campaign was the beginning of the end of the French Emperor and his Empire.

GRAINS ARE THE PERFECT FOOD FOR WAR

As is clear from the historical examples described above, grains have played a major role in warfare. In fact, both in war and in peace, throughout history most agricultural societies obtained the majority of their calories from one of the three main grain crops in the world, wheat, rice and corn (aka Indian corn, or maize), as well as from some closely related species such as barley, rye, oat, millet and sorghum. Even today, the grains of wheat, corn, and rice provide 60% of the calories for all people on earth. This number is of course an average. Poor people eat more grain, while more affluent people are able to consume more meat and diversify their plant-derived food as well. Given the major contribution of grains to the human diet, it is not surprising that grain fields and grain stocks become targets for destruction during war. But why have grains turned out to be so important to human survival and success, and why do they become even more so during war?

To answer this question, we need to learn more about the biological properties of grains. But before we discuss how grain crops grow and what advantages they have as a source of food for people, it is useful to understand why plants in general are a good source of food for people. We start by noting that since plants and humans, like all living organisms, all evolved from a common ancestor, they all share very similar genetic and biochemical blueprints. Both humans and plants have genes that are made from nucleic acid. Plant and animal genes use the same genetic code to specify the synthesis of numerous protein enzymes (see Glossary) made from the same types of amino acid building blocks. Our cells all have membranes that are made from similar mixtures of lipids, and both plant and animal generate energy by burning the same carbohydrate, glucose, or another interchangeable sugar.

As discussed in Chapter 1, the main difference between plants and animals is that to run the metabolism of the body and to make new carbon-containing sugar molecules, plants use energy from the sun and carbon dioxide from the atmosphere, whereas animals cannot do that. Plants are also capable of converting the newly made sugar molecules into all other compounds, such as amino acids, lipids, etc., using some of the sugar molecules as the backbone for the new compounds and some of the sugar molecules as fuel, burning them to obtain the energy to synthesize complex compounds. Animals are not able to make sugar compounds from carbon dioxide and light energy, so they have evolved to obtain their sugar fuel by eating plants or eating other animals that eat plants. Animals, including humans, are also capable of using sugars to synthesize some, but not all, of the more complex molecules. Since animals have been eating plants for so long and obtaining not just sugar but all kinds of other chemicals this way, it is not surprising that over long evolutionary time humans and other animals have lost the ability to synthesize some of the more complex chemicals. For example, humans are unable to synthesize anew 10 of the 20 amino acids that are used to make our proteins. Since plants can synthesize all 20 amino acids,[12] we therefore must directly eat plants, or other animals that ultimately obtained these amino acids from plants, to have complete nutrition.

Given the need of animals to eat plants, what are the best plants to eat, and what are the most edible parts of a plant? Here it is important to note that while the initial products of photosynthesis are the simple monosaccharide sugars fructose and glucose, which the plant cells can easily interconvert, these two sugars, and also the disaccharide sucrose, which plants make by combining the two, do not usually accumulate in most plants (with some exceptions, see sugarcane in Chapter 3). Instead, plants use these simple sugars to build more complex forms of sugars, some of which are used to construct parts of the plant body and some as storage forms of sugar to be broken down later and reused. The complex sugars are made of many units of monosaccharide sugar molecules chemically attached to each other, and are therefore called polysaccharides. Two such important polysaccharides are cellulose and starch (Figure 2.1).

Usually, the bulk of the monosaccharide sugars in plants is converted to the polysaccharide cellulose, which is made by linking glucose molecules in a linear fashion. Cellulose is the major component of the wall of plant cells (and the major ingredient of paper and cotton cloths). Because of the particular way in which the glucose molecules are linked to each other in cellulose, which for short here will be described as a "β linkage" (Figure 2.1), once cellulose is synthesized by plants, the plant cells cannot break it down to the glucose units. Furthermore, cellulose cannot be broken down to glucose by humans either, so it has no dietary value to us, unlike glucose, fructose, or sucrose, which we can easily metabolize. Indeed, no animal is capable of breaking down cellulose back to glucose with its own enzymes. If we ate a leaf or a blade of grass, we would get very little useful sugar out of it, and very little of other useful chemicals such as amino acids too, since leaves and grass consist mostly

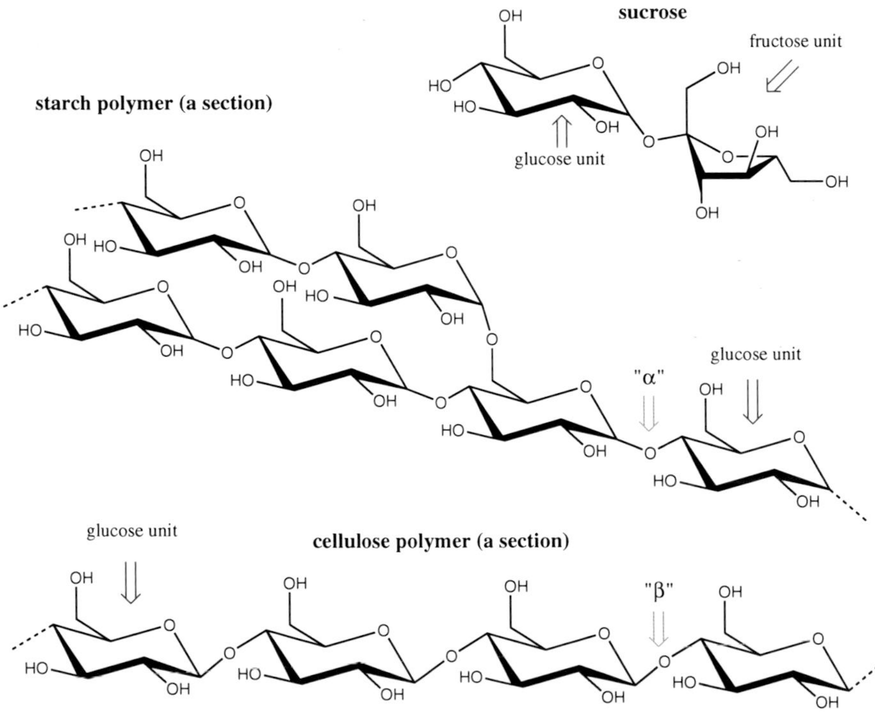

FIGURE 2.1 The disaccharide sucrose, made of glucose and fructose, and the polysaccharides starch and cellulose, each made up of multiple glucose molecules. Both starch and cellulose are linear polymers, but in starch, unlike in cellulose, the linear polymers can also be linked to each other, forming "branches." The green arrows point at the two different ways in which the glucose molecules are linked to each other; an α-linkage in starch, and a β-linkage in cellulose (see Appendix for an explanation of chemical notations).

of cellulose and water. However, some animals such a sheep and cows are able to subsist on such a diet because they have bacteria in their guts that have enzymes that break down the cellulose to glucose, and some of this glucose is released by the bacteria into the gut and is taken up by the digestive system of the animal. Even these herbivorous animals, however, still have to eat large amounts of fresh plant material to survive, and many plants have evolved the ability to synthesize toxic compounds in their leaves as a defense against herbivory (see Chapters 4 and 5).

Unlike cellulose, starch is a glucose polymer that plants synthesize to store their sugars for later use. In starch, the glucose molecules are linked to each other in a different way, called an "α linkage," than they are in cellulose, and plants as well as animals (including humans) have enzymes, called amylases, that can break it down back to glucose. However, plants do not usually store starch in their leaves, but typically in underground structures such as rhizomes, tubers, and bulbs, which are less accessible to herbivores. And there too, these underground plant parts may produce toxic chemicals to protect them from marauding animals. For example, the tubers of wild potato plants contain a toxic chemical called solanine,[13] and the Mexican yam tubers contain the toxic compound diosgenin, both of which have structures similar to animal steroids. Their consumption could interfere with normal steroid metabolism and is therefore harmful. Another example of a starchy tuber is that of the plant *Manihot esculenta*, known as cassava or manioc, native to South America. This tuber is a staple food in Africa, but because of the presence of a compound that decomposes spontaneously to produce cyanide when the tuber is ground into flour, the starchy flour has to undergo a lengthy process of detoxification before it can be consumed. A small amount of cyanide, produced from the grinding of just a few grams of cassava, can easily kill a person.

There is one situation where the plant must deposit and concentrate valuable nutritional chemicals above ground – when it grows its embryos. As described in Chapter 1, plants can reproduce sexually by having an egg cell, present inside the ovule part of the flower, fused with a sperm cell, carried out inside a pollen grain, which is produced in the anther part of the flower. Once the egg and the sperm fuse, the fertilized egg starts growing and dividing and a plant embryo begins to form. Typically, the embryo stops growing after a while and becomes dormant, a condition in which cells stop growing and dividing and the metabolic rate is lowered to the absolute minimum necessary to keep the cells alive. At this point, the embryo and its immediate surrounding constitute a seed, which would usually need to be placed in soil and imbibe water in order to start growing into a seedling (a process called "germination"). The seed is covered by a seed coat, and in some cases it is found inside a structure that we call a fruit. As the seeds and the fruit form on the plant and are physically connected to it, the plant deposits nutrients on the inside of the seed coat and sometimes also on the outside the seed – that is, in the fruit – but these two food deposits serve different functions.

The nutrient deposits in the fruit may include chemicals that are useful to animals, such as sugars, oils, amino acids, and vitamins. The ripe fruit may be brightly colored with pigments, and it often emits an attractive smell (at least attractive to certain species of animals), thus calling attention to itself and inducing specific animals to consume it, together with the seeds inside it. Animals learn from experience to associate specific fruits, which they recognize by shape, color, and smell, with specific nutritional rewards, and will return again and again to the plant that sports these fruits. Once the fruit is consumed, the seeds, protected by their seed coat, then pass through the digestive system of the animal without being damaged, and when the animal defecates, usually at some distance from where the plant that provided the fruit is located, seeds dispersal is thus accomplished. In nature, the process of natural selection has led plants to supply the fruit with just the minimum amount of nutrients that is necessary to make it attractive to the animal and thus accomplish the job of seed dispersal. This is why fruits of non-cultivated plants are relatively small and constitute dilute sources of nutrients (the grotesque size of the fruits of cultivated plants and the high concentration of sugars in them are the result of human selection).

Inside the seed coat, however, the plant deposits as much of the essential nutrients that are necessary for the embryo to germinate and start growing roots, shoots, and leaves before setting up its photosynthesis apparatus that will allow it to start making its own nutrients from carbon dioxide and sunlight. These nutrients that the germinated embryos – the seedlings – need for their pre-photosynthesis growth, these nutrients that the plant therefore supplies the seeds with, are in principle the same as those that people need to consume throughout their lives. So seeds are the best, most concentrated source of foods that people can obtain, However, it must be remembered that because of the versatility of plant biochemistry, even germinating but still non-photosynthetic seedlings can synthesize some of the more complex compounds from basic storage compounds such as sugars and oils, while humans cannot. Therefore some nutrients, such as various vitamins, that humans must consume may not be present at all, or present in only small amounts, in some dormant, pre-germinating seeds. So a diet based on a single type of seed is usually not sufficient, particularly since different types of seeds display a large range in the amounts and concentrations of specific nutrients.

There is a large variation in the size of seeds. Both a grain of wheat and a coconut "fruit" one sees in the store are, practically speaking, seeds.[14] So is the pecan nut in its shell, the pea in the pod, the apple pip and the olive pit. Clearly, as demonstrated in the above sentences, in our colloquial language we have multiple words to describe these seeds.

Generally speaking, seeds can be divided into two principal types – those in which starch is the main storage compound, and those in which oily material is the main storage compound. For example, the seeds of the grasses – wheat, rice, corn, and several others – contain high levels of starch. Seeds that have high oil content include those of rapeseed (canola), pumpkin, and sunflower. A third class of compounds, amino acids, is also present in all seeds but at variable and usually low (<15%) concentration, with the exception of seeds of some of legume plants, also known as pulses. All seeds also contain small amounts of many minerals and vitamins that are necessary for the proper activity of the enzymatic machinery that carries out metabolism.

Plant oils consist mainly of fatty acids. In fatty acids, most of the carbons are bonded to two hydrogen atoms and two carbon atoms, and in sugars, most carbons are bonded to one hydrogen, one oxygen (which is also bound to a hydrogen), and

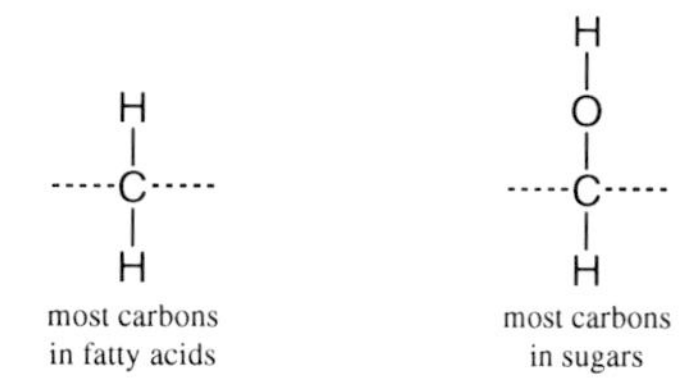

FIGURE 2.2 Each internal carbon in a fatty acid molecule is linked to two hydrogen atoms, and each internal carbon in a sugar molecule is linked to one hydrogen atom and one oxygen molecule. The dashed lines indicate linkage to other carbon atoms.

two carbon atoms (Figure 2.2). The carbons in fatty acids are said to be less oxidized than the carbons in sugar, since they are bound to zero oxygen atoms while the carbon atoms in sugar are bound to one oxygen atom. Chemically speaking, the oxidation of an atom – in our case, a carbon atom – is the process of combining it with an atom of oxygen, and it is also called burning. Burning results in the release of energy, and it can happen both inside and outside living cells. Since carbon atoms can combine with more than one oxygen atom, the fewer oxygen atoms they are bound to initially, the more oxygen atoms they can subsequently bind and thus more energy can be released. This is why burning fatty acids release about twice the amount of energy than burning sugars. When this oxidation process occurs in the living cells under controlled conditions, with the help of enzymes, the cell is able to harvest the released energy and use it later. The process of oxidizing fatty acids in the cell can be total so that the end products are water and carbon dioxide, but under specific, controlled conditions the cell can partially oxidize fatty acids to produce, in a rather circuitous pathway, sugar molecules. The cells of both animals and plants can also reverse the process and reduce (the term means the opposite of oxidize) the carbons in the sugar by removing the oxygen atoms, converting the molecule back to fatty acid, but this requires an investment of energy on the part of the living cell.

As oils and sugars can be easily interconverted by cellular enzymes, why have some plants evolved to endow their seeds with starch, and others with oil? This question is a complicated one that cannot be easily or summarily answered here. But since the three grains that have been most widely adapted by human societies around the globe and that supply most of the energy to most people today – wheat, rice, and corn – contain predominantly starch (Table 2.1), we need to address a related question: Why have starchy seeds proved to be more useful to people than oily ones? This question is particularly interesting because, as also pointed out above, oil contains twice the amount of extractable energy per gram as does starch.

The main answer to this riddle is that starchy grains are much more durable in storage. Oil gets rancid easily – in chemical terms, it gets oxidized by oxygen in the air, and once rancid it is not suitable for human consumption. Starchy grains, kept in dry storage to prevent germination, keep much longer. And once ground into flour, shelf-life can be extended even longer. In fact, starchy grains are the ideal food storage system. As food they do not spoil, they provide most of the nutrients that people need in a very concentrated form, and, being seeds, they can always be sown in the field to make more of them. They hold the advantage both nutritionally

TABLE 2.1
Percent (By Weight) of Different Dietary Components in the Seeds of Wheat, Brown Rice, Maize, and, for Comparison, Dry Soybean[a]

	Wheat	Brown Rice[b]	Maize	Soybean (Dry)
Digestible Carbohydrates	72% (mostly starch)	78% (mostly starch)	75% (mostly starch)	10% (mostly sucrose and other di- and trisaccharides)
Fat	1.5%	2.9%	4.7%	20%
Protein	12.6%	7.9%	9.4%	36%

Note: The remaining material consists of minerals, vitamins, water, and other indigestible organic material.

[a] Nutritional information could be found at https://ndb.nal.usda.gov/ndb/search/list.

[b] White rice, a modern invention, is obtained by removing the seed coat and is highly inferior in nutritional value to brown rice. White rice contains mostly starch.

and in durability over seeds with high oil or protein content because the latter two either spoil easily or contain a less-than-ideal ratio of carbohydrate to amino acids. They also have four times the concentration of starch than potato tubers, which have a much higher content of water and therefore rot easily (because the high water content allows microbial growth). In short, seeds in general are a better source of food compared to fruits or tubers, which only have low concentrations of nutrients and easily spoil unless laboriously preserved. They are overall also a superior source of food to meat, which must be painstakingly preserved and, once prepared through the process of slaughtering, cannot be propagated as can be done with seeds. And of all the type of seeds, starchy grains optimally combine all these advantages.

It is interesting that wheat, rice and corn, the three main grains used by people around the world, all belong to the same family of plants, Poaceae, commonly known as the grasses, even though they were domesticated in very different parts of the world. Wheat was domesticated in the Middle East 10,000 years ago. Domesticated wheat is a species that came into being via a series of hybridization events, first between two ancestral species, then between the resulting hybrid and a third ancestor, and consequently the most commonly grown wheat species, *Triticum aestivium* (Figure 2.3), contains genes from three different species. It is still not clear if the entire process occurred naturally or was aided by ancient humans. Some of the ancestral and hybrid species can still be found growing naturally in the area of the Middle East called the Fertile Crescent, which includes Iraq, Syria, and Eastern Turkey, and one of those, the hybrid emmer wheat (also called Farro or durum) is still being grown commercially for special purpose flour, for example, for making pasta.

A hallmark of wheat is its ability to develop tillers, which are side branches. The main branch and the side branches each eventually develop a "head" with multiple flowers (called "florets") that are pollinated by wind transport of the pollen. Depending on the fertility rate of the florets, after pollination typically 25–40 fruits, or grains, will develop on each "head." This means that for each seed put in the ground, the farmer could reap more than a hundredfold return.[15] In general, wheat

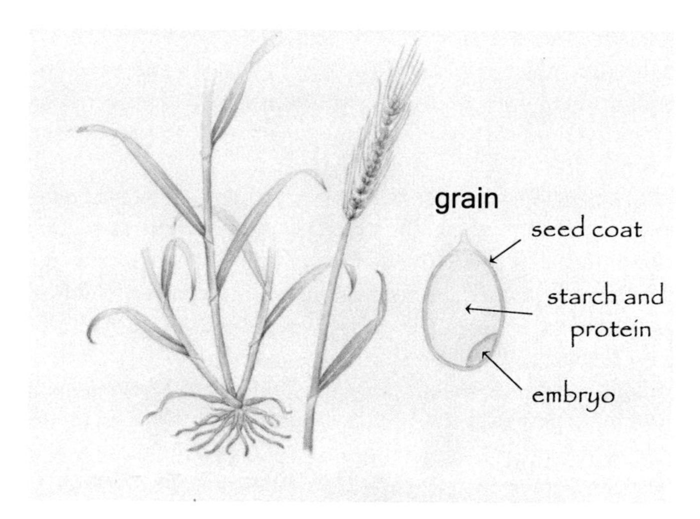

FIGURE 2.3 Bread wheat (*Triticum aestivium*) and a schematic diagram of its grain (seed).

requires moderate amounts of water and therefore it can grow without irrigation even in areas with limited rainfall, as long as the precipitation occurs during the short growing period of the wheat plant (four to eight months, depending on the time of the year that the wheat seeds are sown). It is also able to grow well in a wide range of temperatures, so its cultivation quickly spread throughout Asia and Europe, and, after connections between the Old World and the New World were established in the 16th century, also into the New World.

The domestication of rice (scientific name: *Oryza sativa*) (Figure 2.4) occurred in China, perhaps a thousand years after the domestication of wheat in the Fertile Crescent. Rice, like wheat, also develops tillers that bear heads of grain. However,

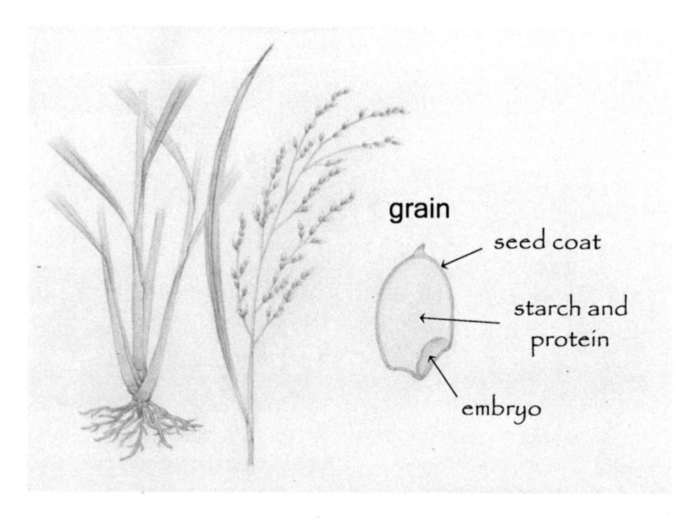

FIGURE 2.4 Rice (*Oryza sativa*) and a schematic diagram of its grain (seed).

rice requires much more water – it does best when grown in a flooded field, or "paddy", and the vast majority of rice is grown this way – and in relatively warmer temperatures than wheat does, so its cultivation has been more geographically limited. Paddy rice also requires twice the amount of work to prepare the field, plant and tend the plants, drain the field before harvesting, and finally harvest the grains. However, yields per acre are about twice those of wheat. Consequently, its cultivation has now spread all over the Far East of Asia as well as to India, Africa,[16] Southern Europe, and, since the 16th century, to the Americas. It is estimated that today 11% of all arable land in the world is used for rice cultivation, and that rice is the primary source of calories to two billion people, more than a quarter of all people living on Earth.

The domestication of the species *Zea mays* (Figure 2.5) occurred in Central America approximately 5,000 years ago, considerably later than that of wheat and rice. We call this species "corn," but this word in English was originally used to denote any grain grown by local farmers. When the English settlers in North Americas first encountered this plant, they called it Indian corn, and the word was later shortened to simply "corn." Corn is an unusual grass species. It has separate male flowers, called tassels, at the tip of the main stem, and female flowers, or "ears," that develop off the main stem just above the base of a leaf. How corn evolved from its ancestor, believed to be similar to an extant plant called teosinte that is found in Mexico and Central America and that has only 6–10 kernels in each ear, is somewhat of a mystery. However, with so many corn cobs of intermediate size and of different archeological age, it is clear that it took many thousands of years of human cultivation and selection for the corn cob to achieve the size and the number of grains that a present-day corn cob has.[17] Modern corn hybrids yield as much as 9,000 pounds per acre, slightly more than the yield of paddy rice, and two and a half times more than the average yield of a wheat field. The cultivation of corn, also a plant that requires warm temperatures to grow but only moderate amounts of rainfall, was initially restricted to Central America and the temperate zones of the North and South

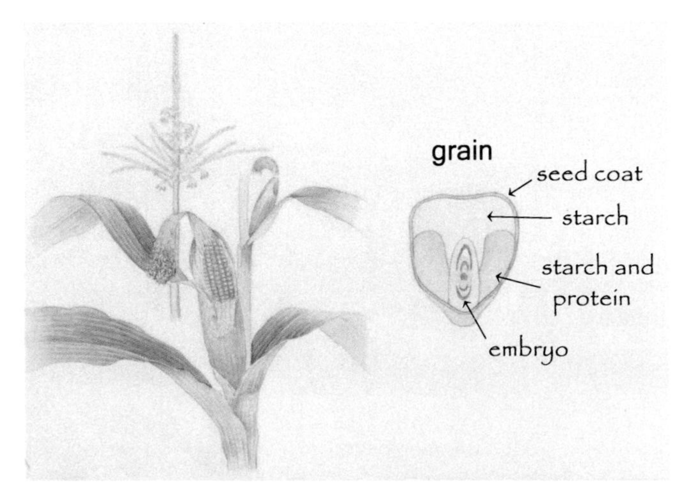

FIGURE 2.5 Corn (*Zea mays*) and a schematic diagram of its grain (seed).

Americas. However, since the 16th century, its cultivation has spread to all settled continents of the world.

Being close relatives, the grains of these three plant species have much in common. Their starch content is very similar, approximately 72%–80% of the total weight of the seed. The grains also contain a good concentration of protein, with wheat having the highest concentration, approximately 13%, and rice having the lowest, approximately 8%. The animal digestive system breaks down the proteins to their amino acid components, and it turns out that corn proteins have low content of the amino acid methionine, one of the "essential" amino acid that humans must acquire in their food. Corn seeds also have the highest concentration of oil (4.7%) compared to rice and wheat, and typically a higher level of moisture (16% for corn compared to 11% for wheat), and corn kernels are therefore much more likely to spoil during a long storage period. For these and other reasons, today most of the harvested corn is not consumed directly by humans but is used as animal feed and in the production of high-fructose sweetener and ethanol fuel.

Nonetheless, all three of these grains constitute excellent, concentrated food sources. All three can also be ground to produce flour, which is then mixed with water to make dough and baked to obtain easily digestible food. Rice and corn flours lack the gluten proteins, which give dough made from wheat flour its elasticity and thus allow it to rise when first incubated with yeast or, alternatively, mixed with chemical ingredients that generate carbon dioxide gas, such as baking soda. Baked goods made from the flour of any of these three grains represent a high-energy and dense food source, since most of the water evaporates during the baking process. The grains of all three plants can also be cooked directly in water and eaten right away or after further processing.

It is therefore not surprising that, since antiquity, armies embarking on a campaign made sure that they took sufficient amounts of grain with them. It has been estimated that Persian, Greek, and Roman soldiers consumed three pounds of grain a day, which supplied a bit more than 60% of their caloric needs.[18] When time allowed, the grain would be ground and used to bake bread, but when time was short the grain could simply be boiled in water and eaten as porridge. Even today, with modern technology allowing the efficient and safe preservation and packaging of meats and milk products, soldiers' rations contain a fair amount of hardtack biscuits[19] and dense bread and cookies.

Provisioning the troops was a major undertaking. Living off the land was often not an option when local resources were limited in relationship to the size of the army. The Persians employed 3,000 transport ships to sustain their army during the invasion of Greece in 481 BC, with most of these ships carrying grain and other foods.[20] On land, ancient military logistics employed carts pulled by oxen, horses or donkeys. However, such transport was slow and itself consumed a fair amount of food – by the animal, by the drivers, and by the repair crews. Thus, for gaining speed, particularly in expeditionary campaigns, since the time of King Philip II of Macedonia, father of Alexander the Great, the foot soldiers of the Greeks, the Romans, and all other armies that followed have been required to carry their own food (as well as their weapons and equipment). With the limited carrying capacity of the human body, grains constituted most of the food being carried, so when the soldier was unable to find additional food locally, his diet consisted mostly of grain. Thus, the durable grains, being a source of highly concentrated calories, enabled quick and long-distance military campaigns. In the absence of such grain or an efficient transport system, on the other hand, such military operations could not be carried out; for example, in Central America, Mayan warfare occurred mostly between cities that were separated by no more than a 2–3 days' walking distance.[21]

GRAINS AND MAN-MADE FAMINES IN THE 20TH CENTURY

As we have seen, since most people rely on grains for most of their caloric needs, the absence of grain during war often leads to mass starvation. But in the 20th century, major famines have occurred in some countries while these countries were not involved in wars with either external adversaries or even in civil wars. On the face of it, with the world being so interconnect by then, such famines should not have happened, since help with food supplies could have been shipped from other countries. But several major famines that killed many millions of people did occur, and they all had two things in common: they were largely caused by human action and they happened in totalitarian countries with collectivist (i.e., non-market) economies. Four such famines, accounting for more than 60% of all famine deaths in this century, occurred in the Soviet Russia in 1932–1933, Communist China in 1959–1961, Cambodia under the Khmer Rouge regime in 1975–1979, and Communist North Korea in the 1990s. In the Russian and Cambodian famines, there is historical evidence that the political leadership responsible for these famines intended to cause at least some deaths and that both direct deaths from starvation as well as violent deaths were non-proportionally visited upon specific groups of people in the population that the leaders found "undesirable." The Chinese famine, while not strictly intentionally caused, nevertheless indicated depraved indifference on the part of the leaders to the lives of certain groups among the population, and was also accompanied by violent deaths perpetrated by the agents of the regime. In essence, the leaders of these countries engaged in a war against their own citizens. The specific causes of these three famines, which have a lot in common but also some interesting differences, are described below (there is not enough historical research at this point to discuss the exact causes of the recent North Korean famine).

THE RUSSIAN FAMINE OF 1932–1933[22]

When the Communist Party took control over the territories of the former Russian Empire in October 1917, the Party under Lenin began a gradual if uneven process of converting the economy to one controlled by the State, as opposed to a market economy where each person is mostly free to act according to what he perceives to be his best interest. This process involved collectivization, in which the state assumed ownership of all property and the citizens simply worked for the State, and private property rights were abolished. In the early 1920s, Lenin, the head of the Soviet Union, was forced to retreat for a while from establishing a true communist economy in the countryside in the face of fierce resistance from farmers. Following Lenin's death in 1924, Stalin took his place, and, after a few years of consolidating his control over the Communist Party, he resumed the farm collectivization process in 1928. Simultaneously, he pushed for industrialization of the still mostly agricultural economy of the Soviet Union.

The Russian Empire has always been heavily dependent on grain produced in a southern "belt" stretching from the Ukraine in the west to Kazakhstan in the east. As the farm collectivization policy was being implemented in fits and starts, annual goals for agricultural production were determined by Soviet functionaries as part of the first "five-year plan" approved by Stalin himself and his group of top communist leaders (the "Politburo") in 1928. And every year, a portion of the planned grain harvest, to be produced on the collective farms (kolkhozes), was slated to be taken by the state to feed urban populations and to sell abroad to obtain foreign currency to buy equipment for industrialization.

However, despite the planned increases, grain yields following collectivization actually began to decrease because fewer people were allowed to remain on the farm and because under the communist system individuals had no incentive to work hard or efficiently since they were not preferentially rewarded for doing so. In fact, they often fell under suspicion for being too enterprising, thus betraying a non-socialist mindset. So, although the communists viewed all farmers in general as reactionary and unsympathetic to the communist cause, they showed special animosity toward the more successful farmers, whom the Communist Party called "Kulaks." This was a politically loaded term that was applied somewhat indiscriminately to anybody who showed any entrepreneurial initiative and in particular to successful farmers with large land holdings who hired non-family farm laborers to work on their farms, thus "exploiting" these workers' labor according to communist terminology. However, it was exactly such farmers that were responsible for most of the grain production, and these were declared class enemies and relentlessly persecuted.

Given the independent streak of farmers in general, it was not surprising that most farmers resisted collectivization. The Resistance was sometimes violent – attacking government officials – but more often passive, such as refusing to work efficiently or hiding food, and it was particularly strong in the Ukraine, where it was also motivated by the desire for national independence.[23] The Soviet government, for its part, responded with strong violent repression, again particularly harshly in the Ukraine, with large numbers of people arrested and deported to Siberia (where many died), or executed outright.

With farming activities diminished because of collectivization, and the state continuously ramping up its extraction of grain from the grain-producing areas, finally in 1932–1933 a major famine developed in the grain-producing areas (but not in non-producing areas) in which at least as many as ten million people died. There was simply not enough grain left for the farmers to eat and to keep grain for sowing next year. Deaths were particularly high in the Ukraine, with some estimates indicating as high as seven million dead. There is evidence that, in August 1932, Stalin himself gave orders to increase the seizure of grain, being harshest on the Ukraine, in order to cause mass starvation and thereby force the local populations to submit to collectivization and to give up their aspiration of independence from the Soviet Union. Starvation indeed became apparent in late 1932 and intensified in early 1933. Many people, weakened by malnutrition, succumbed to diseases such as typhoid fever and malaria. To compound the problem, the Soviet officials, who generally mistrusted the population and believed that people horded food, sent the police to search houses and remove any food that they found, again targeting Kulaks and other people thought to be hostile to the Soviet regime. At the same time, the authorities forcibly prevented rural residents from leaving the farms and moving into the cities, where food was generally more plentiful. News reports of the famine were suppressed and combatted with the deliberate release of misinformation,[24] and unlike in an earlier, smaller-scale famine that occurred in the same grain-producing areas during the civil war of 1918–1921 (also because of excessive grain requisition by the Bolsheviks), no foreign aid was allowed to reach the afflicted areas. In late 1933, after millions had died, Stalin allowed more grain to remain in the rural areas and the famine subsided.

THE CHINESE FAMINE OF 1959–1961[25]

About 30 years later, between 1959–1961, a famine occurred in Communist China that killed an estimated 20–45 million people (most recent estimates fall in the upper part of this range). In terms of deaths, it was probably the worst famine in human history. And it occurred for many of the same reasons that caused the Soviet Famine of 1932–1933, although for the most part apparently without overt murderous intentions on the part of the leadership – only depraved indifference to the loss of life.

The Communists, with Mao as their leader, gained control of China in 1949. As communist economic orthodoxy mandated, the communist leadership promulgated five-year economic programs that the country must strictly follow. In 1958, Mao decreed a new five-year plan titled the "Great Leap Forward," which was intended to complete the collectivization of the country's mostly agrarian economy and to develop its industry so that it would catch up with the Western capitalist economies. The Great Leap Forward called for the completion of privatization of land ownership, started already in 1949, and the transfer of all farmers to communes. An important feature of the communes was the communal dining hall where the farmers stored, prepared, and ate their food, with no food allowed to be kept in individual abodes.

That five-year plan caused a tremendous famine, unprecedented in human history. While the Soviet famine was caused mostly by removal of grain from rural areas and not because of a major decrease in grain production (although some decreases did occur), the situation in China between 1959–1961 was more complex. In China, rice is grown mostly in the south and wheat, as well as some sorghum and corn, are grown in the north. During this great Chinese famine, there were indeed major decreases in grain production, with production in 1959 down 15% compared to the year before, down another 15% in 1960, and another decrease of 5% in 1961, the year with the lowest grain yields.

Some of the decreases in production were indeed due to bad weather, but probably only a small fraction. Even by official Chinese estimates, natural causes accounted to no more than 30% of the decrease, and the actual percentage is probably still smaller. Instead, most of the decrease was due to several man-made causes. First, as in the Soviet famine of 1931–1932, there was a general lack of enthusiasm for collectivism among farmers. Also, many rural residents were diverted to the local production of steel – an obsession of Mao – and many agricultural implements were sacrificed to make what ended up as useless pig iron, a form of unrefined iron that contains too much carbon and is therefore brittle and useless for the construction of machinery.

But the largest contribution to the development of the famine was made by bad agricultural practices mandated by ignorant communist operatives following the suggestion of the advisors that the Soviet Union sent to China to help out their communist brethren. In the Soviet Union at that time, the charlatan Trofim Lysenko held sway as the foremost "geneticist" of his time and an expert on agriculture.[26] Among the misguided agricultural methods that Lysenko advocated, and that the Soviet advisors in China therefore promoted, were extremely deep plowing (as much as a meter deep, three times more than is normally done) and sowing grains at very high density, claiming both practices will increase yield. However, the deep plowing caused minerals to come up to the surface, increasing the salinity concentration in the soil to a point that harmed the plants, and the high density of the wheat and rice plants simply suppressed the development of tillers, so no increase in yield per acre was achieved, with much wastage of seeds. Another ill-advised policy that contributed to the famine, this one devised by the Chinese political echelon on their own, was the extermination of the Eurasian Tree Sparrow because they believed it ate grain. It led to a huge outbreak of insect pests (which were the sparrow's real food) in the field, and subsequently to great loss of crop.

But as with Russia, the direct cause of the famine and the huge death toll was the forced removal of grain from rural areas, combined with political persecution. The Chinese Communist Party was controlled by Mao through the development of a cult of personality in which his decisions were worshipped as infallible. Those who did criticize him were severely punished. The Great Leap Forward was promulgated by him and it called for, among other things, increasing grain production to feed a growing urban population (the "proletariat") and to sell abroad to generate funds for industrial machinery and to pay back international debts. The wisdom of his decisions could not be questioned, and if their outcome was not positive, it must be because some people were engaged in active sabotage. The state extracted as much as 30% of the grain produced in each area, with the amounts determined based on last year's yields. These yields were often exaggerated by local officials trying to curry favor with the central leadership, so a higher proportion of the grain was actually removed than was intended by the central administration, exacerbating the famine. On the other hand, local officials often suspected that farmers horded food, and armed local political cadres regularly hounded individuals, torturing and killing many for perceived offenses against the communal enterprise. By some estimates, as many as 10% of the total victims died as the result of violence. Finally, there is good evidence that in each locale, remaining food was unevenly distributed, with politically disfavored individuals thus suffering disproportionally. And as with most famines, the very young and very old suffered the highest rates of mortality.

Truthful reporting up the chain of command was considered, for good reasons, dangerous by low-level functionaries, and even high-level officials were afraid to report the ongoing famine to Mao. However, Defense Minister Peng Dehuai did write a personal letter to the Chairman with detailed information in 1959, and as a consequence Mao deposed him from his job. Mao is quoted as saying in a secret Shanghai meeting in 1959 that "when there is not enough to eat, people starve to death. It is better to let half the people die so that others can eat their fill." At that meeting, he insisted on the target of 33% procurement of grain, saying that "if you do not go above a third, people will not rebel." It took until 1961 for some people in the Communist leadership to band together and sideline Mao from his control of the party. The policies of the Great Leap Forward were then discontinued, grain reserves were sent back to rural areas, and grain levies were reduced drastically, ending the famine. China also began to import large amounts of grain from capitalist countries such as Canada and Australia.

THE CAMBODIAN FAMINE AND GENOCIDE OF 1975–1979[27]

There was nothing unintentional about the genocide that occurred in Cambodia between 1975–1979, in which at least two million people – 25% of the population – died from starvation, disease, and violent acts ordered by the communist leadership of the country.

Cambodia began the 20th century as a colony of France. It became independent in 1953 but remained embroiled in the regional conflict centered around Vietnam. Starting in the late 1960s, Saloth Sar (known as Pol Pot) began to take over the Cambodian Communist Party (known as the Khmer Rouge) and to use its military force to gradually take control of the country, an effort that culminated in the capture of the capital Phnom Penh by the Khmer Rouge in April 1975. Once in control of the whole country, the Khmer Rouge embarked on a project to turn the entire country into an agrarian, non-capitalist utopia. The crash program that Pol Pot and his comrades, who were inspired and militarily supported.by China, had in mind was a combination of the Chinese Great Leap Forward and the Cultural Revolution, the latter a program that Mao initiated in the mid-1960s to fight "class enemies" of the proletariat (and to regain his control of China). For Pol Pot and the Khmer Rouge in Cambodia in 1975, everyone who was not already a farmer – their version of the proletariat – was considered a class enemy, as well as many farmers too, those who had not signed up to the Khmer Rouge's utopian (or dystopian) vision. So once the Khmer Rouge were in total control of the country, all the residents of cities were forcibly removed to the countryside and organized into working brigades. This was done to accomplish two main goals. The first was to grow rice for export to obtain money to buy other necessities (including military equipment). The second was to kill all the class enemies, whom the Khmer Rouge believed could never be turned into their idealized peasants, through hard work, starvation, torture, and execution.

When the Khmer Rouge took over the country, they closed the country to outsiders, money and all private commerce were abolished, and all social and religious organizations were banned. Any manifestations of modern life were eliminated, including medicine, education, and electronic communication methods, except when they helped the interests of the Khmer Rouge leadership. The population became slaves in their own country, working 18 hours a day under armed guards, and food was rationed to one cup of rice per day (approximately 180 grams, or 680 calories). Any minor infraction of the rules could result in immediate execution or in slower death by torture. The regime periodically carried out large-scale massacres of groups of "old society undesirables," of those belonging to non-Khmer ethnic groups, or of workers who were no longer needed or who could not perform their assigned duties because of weakness due to starvation and disease. Many Khmer Rouge soldiers and political cadres whose loyalty became suspect were also executed. It is estimated that at least half of the total deaths during that period were caused by violent means, the rest due to starvation and disease.

Incredibly, Pol Pot and the other addled Khmer Rouge leaders, while busy exterminating their own people, also found time to attack neighboring Communist Vietnam. After repeated military incursions by Khmer Rouge forces across the international border into Vietnam, the Vietnamese army invaded Cambodia in December 1978 and quickly occupied most of it. The Vietnamese occupation brought the murderous reign of the Khmer Rouge to an end in all but a small, forested part of the country in the west near Thailand, where remnants of the Khmer Rouge held control for about 20 more years, being supported there by China, Thailand, and, tacitly, the United States.

CONCLUSION

Since the domestication of wheat, rice, and corn, their grains have provided the majority of the calories for the survival and growth of the human species. The success of grain agriculture, with some help from other types of crops and domestic animals, led to large increases in human populations, to urbanization, social stratification, and then to the establishment of armies and the frequent occurrence of warfare among neighboring human societies. The advantage of grains as an easily transportable, concentrated form of food has also directly contributed to human aggressiveness by enabling mobile warfare. Invading armies have been fueled by grain; it is grain that allows an army to move quickly and wrest control of large swathes of territory from an adversary. It is not a coincidence that the biggest empires in the world were built by societies that were particularly successful at growing and controlling grains. And because of the crucial importance of grain for the performance of the troops both in preparation for war and during war, much of the action in warfare has involved combatants trying to obtain grain to feed their own side and to prevent the other side from having any.

The reliance of humans on grains as their major source of calories has also put civilian populations at great risk when shortages occur during war with external enemies. However, history shows that the greatest death tolls due to starvation of civilian populations have been caused by domestic disputes. Such domestic disputes in which starvation was used as a weapon include the Biafra secession war in the early 1970s (see Chapter 8), and more recently, the civil war in Syria in the 2010s. But most lethal of all have turned out to be wars waged by totalitarian regimes against segments of the population under their control with the goal of subduing and sometimes simply eliminating those who oppose the regime's control of the country and its resources.

NOTES

1. Jared Diamond's book *Guns, Germs, and Steel* (see Note 1 in Chapter 1) provides a detailed description of how agriculture originated and developed multiple times around the globe and why specific grasses proved to be better at being domesticated than other types of grasses.

2. We also need oxygen for burning the fuel, and water in which the chemical reactions could occur and that also helps dissipate the extra heat produced from burning fuel. As pointed out in Chapter 1, it is also important to remember that besides serving as fuel, the intake of organic matter is also needed to grow the body in the first place and to replace molecules such as amino acids, DNA components, vitamins, etc., that regularly break down and are discarded by the body.
3. Peel, M. 1997. Hunger strikes. *British Medical Journal* 315: 829–830.
4. Protocol Additional to the Geneva Conventions of 12 August 1949, and relating to the Protection of Victims of International Armed Conflicts (Protocol I), 8 June 1977. Article 54 – Protection of objects indispensable to the survival of the civilian population. https://ihl-databases.icrc.org/ihl/WebART/470-750069?OpenDocument.
5. Officially called the Delian League.
6. Described in Thucydides, *History of the Peloponnesian War*, translated by Rex Warner. 1954, Penguin Books. In general, until very recently, military campaigns in temporal zones of the world tended to be halted during the winter season.
7. The Old Testament contains the following fanciful story:

   And Samson went and caught three hundred foxes, and took firebrands, and turned tail to tail, and put a firebrand in the midst between two tails. And when he had set the brands on fire, he let them go into the standing corn of the Philistines, and burnt up both the shocks, and also the standing corn, with the vineyards and olives (Judges 15, 4–5, King James Bible).

   A firebrand is a torch, and the Hebrew original is clearer about what actually got burnt: sheaves of corn (wheat or barley), standing dry corn in the field, and olive orchards.

   While there is no archeological evidence for the historical veracity of this story, it certainly indicates the familiarity of the writer with the concept of burning grain fields. Sherman's March to the sea is vividly described in Marszalek, J. F. 2014. Scorched Earth – Sherman's March to the Sea. *Hallowed-Ground Magazine*, Fall.
8. The Malay Emergency was not defined as war by the British so that the British plantation owners would not lose insurance coverage for their damages. Their Lloyd insurance policies had an exception for war.
9. Agent Orange is a mixture of two compounds, 2,4,5-T (2,4,5-Trichlorophenoxyacetic acid) and 2,4-D (2,4-Dichlorophenoxyacetic acid) that are similar in structure to a plant hormone called auxin (indole-3-acetic acid), which controls, at extremely low concentrations, many biochemical and physiological processes in the plant. The plant confuses these compounds for the hormone, and when these compounds, which the plants cannot breakdown, are applied to plants in high concentrations, many of these biochemical and physiological processes go awry and the plants eventually die. Auxin itself as well as 2,4,5-T and 2,4-D have no serious adverse health effects on humans; the damage to human health exposed to Agent Orange spraying presumably came from a contaminant, dioxin, found in Agent Orange.
10. From Gibbon, E. 1788. *The Decline and Fall of the Roman Empire.*
11. Cronin, V. 1971. *Napoleon.* HarperCollins Publishers.
12. The ten amino acids that humans cannot synthesize in their bodies are called "essential amino acids," meaning that they must be acquired through nutrition. All 20 major amino acids (and a few rare ones too) are essential to build up the proteins in human cells. For background on the biochemical capabilities of plants, see Heldt, H. W. and Piechulla, B. 2011. *Plant Biochemistry*, 4th ed. Elsevier.
13. But tubers of cultivated potato plants produce very little solanine.
14. To be botanically correct, the grains of wheat, corn, and rice are considered fruits, because on the outside of the seed coat there are a few additional dry layers that are derived from the ovary. The hard coconut shell is also derived from the ovary (two

other layers are typically removed before the coconuts are sold in the store) and is thus a fruit. But for our purpose here – following where the nutrients are – the nutritional value of such "fruits" lies completely in the seed part.

15. Guo, H. W. and Schnurbusch, T. 2015. Genotypic variation of floret fertility in hexaploid wheat revealed by tiller removal. *Journal of Experimental Botany* 66: 5945–5958.
16. A distinct rice species, *Oryza glaberrima*, was independently domesticated in Africa but today *Oryza sativa* is mostly grown there.
17. Doebley, J. 2004. The genetics of maize evolution. *Annual Reviews of Genetics* 38: 37–59.
18. Gabriel, R. A. and Metz, K. S. *A Short History of War and the Evolution of Warfare and Weapons*. Strategic Studies Institute, US Army War College (online version): http://www.au.af.mil/au/awc/awcgate/gabrmetz/gabr0000.htm.
19. Also used by sailors on long voyages in the era prior to the refrigeration of food.
20. Herodotus. *The Histories*, translated by Aubrey De Selincourt/1954, Penguin Books.
21. In South America, the Inca army used freeze-dried potatoes for long-distance campaigns.
22. The historical narrative here follows mostly Conquest, R. 1986. *The Harvest of Sorrow: Soviet Collectivization and the Terror-Famine*. Oxford University Press.
23. Ukraine was recognized as an independent country by the United States until 1934, when President Roosevelt rescinded this recognition and the United States recognized the Soviet Union instead and opened diplomatic relationships with it.
24. Abetted by Alter Duranty, a reporter for *The New York Times* who reported that there was no famine in the Soviet Union and received the Pulitzer Prize in 1932 for his reports. In the 1980s, *The New York Times*, after an internal investigation, requested the Pulitzer committee to rescind the prize, but the committee has refused to do so.
25. The historical narrative here follows mostly Dikötter, F. 2010. *Mao's Great Famine*. Walker & Co.; and Meng, X., Qian, N., and Yared P. 2015. The institutional causes of China's Great Famine, 1959–1961. *Review of Economic Studies* 82: 1568–1611.
26. Trofim Lysenko (1989–1976), a Ukrainian by birth, was a Soviet geneticist and served as the head of the Institute of Genetics in the Russian Academy of Science from 1940 to 1965. Earlier he claimed to have developed agricultural methods that raised the yield of major crop plants. These claims, as well as his political skills in maneuvering within the Soviet system, made him powerful within the Soviet scientific community. He used this influence to silence critics who pointed out that his scientific theories were wrong (for example, he did not accept Mendel's explanation of inheritance as based on genes transmitted from one generation to the next) and that his methods of improving agriculture did not work.
27. The historical narrative here follows mostly Kiernan, B. 1996. *The Pot Pol Regime: Race, Power and Genocide in Cambodia under the Khmer Rouge, 1975–1979*. Yale University Press.

3 War and Slavery Capitalism – Sugarcane, Tobacco, and Cotton

TRANSCONTINENTAL STAPLES

As we have seen in Chapter 2, since the beginning of the development of agriculture grains have constituted the major source of calories in the human diet. Grains allowed human societies to get large and to become urban and sophisticated, with individuals developing specialized roles and with castes forming, including a warrior caste. Land used for growing grains was often in contention and this led to wars between different groups of people. As some societies, such as the Athenian and Roman Empires, extended their borders by warfare and were thereby able to grow grain at a long distance from their urban centers, this necessitated the long-distance transport, mostly by ship, of grain to the center. Other than grains, though, until about the middle of the 15th century no other plant product was transported such long distances in such large quantities as grain. True, wine and oil were also transported by ship to countries bordering the Mediterranean and Black Seas, but at much lower volumes. And the trade in spices from the South and Far East to the Middle East and Europe, both by land and by sea (Chapter 4), was always a case of high-value, low-weight products.

But in the 15th century, as a direct result of the onset of the Age of Discovery that was initiated by the naval expeditions of the European countries in search of spices (see Chapter 4), this situation began to change, and at the center of this change were three plants serving diverse human needs or desires – sugarcane, tobacco, and cotton. Of the products of these three plants, only sugar was initially considered a spice, but soon outgrew this definition. None of them was essential for human survival. Nevertheless these three plants, which originated in different parts of the world, became major commodities for the European, and indeed world, markets. In the process of becoming "essential" to modern people, they were at the center of many wars of expansion and other violent events. And these events have had major effects on the direction of world history and the formation of modern states, with ramifications that continue to be felt to this day.

SUGARCANE

A strong case can be made that sugarcane was the root cause behind the development of the capitalist market economy that prevails in many of the world's countries today.

But before we discuss this idea further, we need to describe why sugar, the chemical that makes sugarcane so important to people, is made by plants in the first place.

As discussed in previous chapters, plants convert carbon dioxide, water, and light energy to monosaccharides such as glucose and fructose. Once plants make monosaccharides, they then use these molecules as fuel for their own metabolism, or as building blocks for growing and making more tissues, or they can store them for later use. As an example of the first, plants link many glucose molecules to form the polymer cellulose, which constitutes the plant cell wall that gives plant cells protection from the elements and pests. Cellulose will be discussed in more detail later in this chapter in the section on cotton.

Another polymeric carbohydrate made by plants and discussed at length in Chapter 2 is starch. Starch is also a polymer of glucose, but in starch the glucose molecules are linked together slightly differently than in cellulose, so plants – and animals – are able to break down starch back to glucose, which they cannot do with cellulose. Plants make and store starch in specialized organs, often in tubers or rhizome underground, but also in their seeds and fruits. To make starch, though, the organs that make it need glucose, which must be delivered from the leaves, where photosynthesis occurred. Most extant plants, and all crop plants, are vascular plants – they have one network of tubes, collectively called xylem, that run through their organs from the roots to stems to leaves and deliver water, and they have another network of tubes, collectively called phloem, that run parallel to the xylem and deliver nutrients, including sugar. While water movement in the xylem is typically from the roots to the aerial parts, movement of solutes in the phloem can go in either direction, depending which part of the plant has higher concentration of them (this part of the plant is called the "source") and which part has the lower concentration ("sink").[1]

It turns out, however, that the sugar molecules that move through the phloem of plants are not glucose or fructose, but the disaccharide sucrose (see Figure 2.1 in Chapter 2), with each sucrose molecule made by combining one molecule of glucose and one molecule of fructose.[2] In general, it is sucrose molecules rather than monosaccharides that move from one cell to another in the plant body (as we have seen, the chemical term "sugar" refers to a class of compounds, but in colloquial use, "sugar" usually means sucrose). Once the sucrose reaches its destination – be it the roots, which need nutrients because they cannot photosynthesize for lack of light, or a new leaf that is not yet capable of photosynthesizing and needs nutrients to grow, or the tuber whose function is to store carbohydrates – the sucrose can then be broken down back to its glucose and fructose components, which can then be used further. It helps that fructose can be easily converted by the plant's enzymes to glucose, and vice versa.

For reasons unknown to us, plants in the genus *Saccharum*, which is in the grass (Poaceae) family – the same as the grains discussed in Chapter 2 – have evolved to simply store their excess carbohydrates in the form of sucrose, and to mostly forgo the conversion to starch for the purpose of storage. The name "sugarcane" has been applied to several species in this genus, but the sugarcane species that is most heavily cultivated today is *S. officinarum*, the so-called "noble cane" (Figure 3.1). This species is a hybrid that came into being by successive cross-breeding events between several species, similarly to the development of cultivated wheat described in Chapter 2, again possibly with the active involvement of humans. Sugarcane was first domesticated in New Guinea about 8,000 years ago, and from there it spread west

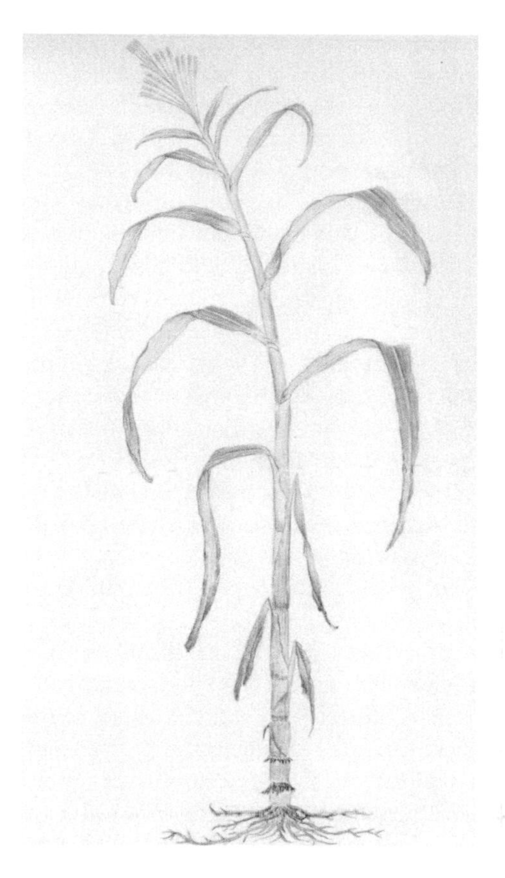

FIGURE 3.1 A sugarcane plant.

to India as well as east to the Pacific islands during their colonization by Polynesian voyagers. Alexander the Great and his soldiers encountered sugarcane when they reached India, and so did the Persians, who began to grow the plant in Mesopotamia. When the Arabs established their empire in the 7th–8th centuries AD, they spread the cultivation of sugarcane all the way to Morocco and Spain in the west as well as to islands in the Mediterranean Sea, including Sicily, Cyprus, and Crete.[3]

Sugarcane is a perennial lowland tropical plant that requires a lot of water and warm temperatures throughout the year.[4] With few exceptions, it does not grow well beyond roughly latitude 30° N and 30° S. As mentioned earlier, sugarcane is one of the few plants that store the carbohydrates (a general chemical term for sugars) they produce by photosynthesis as the disaccharide sucrose (Chapter 2). In sugarcane specifically, the sucrose is stored in the above-ground stem, so a lot of the plant sucrose is easily accessible for harvesting by people. Typically, it takes about 12–18 months from the time small sections of cut stems are planted in the ground until they grow into mature plants whose upright stems are ready for harvesting. Harvesting is done by cutting the upright stems at ground level, after which the plants regrow. With this method, a sugarcane field can be harvested three to eight times before yields start diminishing and it is necessary to clear the field and prepare it for replanting. At harvesting, the concentration of sucrose in the stem could be as high as 12%–15% of dry weight. Harvesting must be done before the plant begins to flower, because when it does the concentration of sugar begins to drop. Once the stems are cut and collected, they are ground and pressed to extract the juice. The juice is collected and processed in several energy-intensive steps that ultimately yield crystals of sucrose of varying purity, depending on the number of steps employed, and also molasses, a highly concentrated solution of sugar and other plant compounds, as a byproduct. Often in the past, sugar of medium purity (darker in color) was obtained locally and then shipped long distances, while additional refinement steps that gave purer, whiter sugar crystals were carried out at the receiving end. And until mechanization was applied in the late 19th century to most sugarcane cultivation and processing steps, obtaining sugar from sugarcane was a highly labor-intensive affair.

As is well known, sucrose is a compound much beloved by humans. The human mouth is equipped with five types of taste detectors, also known as "receptors," to identify what chemicals we ingest. We classify the compounds that bind to these receptors as respectively sweet,[5] sour, bitter, salty, and savory, also called umami. The detectors are present on the tips of nerve cells that connect to the brain, and when they bind a chemical (an explanation on the molecular level of how chemicals bind to receptors will be provided in detail in Chapter 4), the nerve cells in which they are embedded send a message to the brain. This message is interpreted by the brain in a somewhat complicated way that can be crudely summarized as generating a distinct sensation that is classified as either "pleasant" or "unpleasant." A reasonable hypothesis is that these taste detectors, similarly to the scent detectors in our nose discussed in Chapter 4, evolved to protect us from ingesting chemicals that are bad for us and to encourage us to ingest chemicals that are good for us. In respect to the sweet receptors, or taste buds, which detect many mono- and disaccharides, it is almost uniformly reported across human cultures that the sensation obtained from ingesting sugar is a pleasant one. There is also evidence that such a response is inborn and not culturally inculcated.[6]

The intensity of the pleasant sensation generated by the sweet receptors upon ingesting sugars depends both on the type of sugar and the amount. Starch, the main ingredient of grains and many other food staples such as potato, does not activate the sweet receptors by itself. However, because humans have enzymes, called amylases, in their saliva that break down starch to glucose and begin to do so as soon as food is placed in the mouth, eating starchy food feels like eating something sweet, but not very sweet. Ingesting foods containing glucose, fructose, or sucrose elicits a much stronger sweet sensation than eating starchy food, given that equal amounts of the sugars are consumed, with fructose being a bit sweeter than sucrose, and glucose a bit less sweet than sucrose as sensed by the average person.

Sugars are highly valuable nutrients to humans, so the ability to detect sugars and the pleasant sensation we obtain from detecting them appear to be useful adaptations, since, once a pleasant sensation has been experienced, we tend to repeat the activity that previously gave us this sensation in order to experience it again, and so we are likely to consume more sugar-containing foods. Leaving aside the issue of overeating and the mechanisms that prevent this from happening (and sometimes fail to prevent it), humans clearly show preference for, and seek out, foods that contain sugars. Until sugarcane came along,[7] though, the only source of concentrated sugar for human consumption was honey, which is a solution of mostly glucose and fructose with some sucrose, that bees collect from flowers. In general, neither animal-derived food nor plant-derived food contains high levels of sweet-tasting sugars. Fruits, of course, often taste sweet, and some types of seeds, for example, some legume seeds, do too. In such cases it is usually due to the presence of sucrose, which is transported to the fruit or the seed from the photosynthetic leaves, and sometimes glucose and fructose, derived from the breakdown of sucrose, are also present. But the concentrations of these sugars in fruits and seeds are rarely as high as the 12%–15% sucrose concentration found in sugarcane stems. Moreover, fruits and seeds, having other valuable nutrients, have never been used to obtain large quantities of pure sugar.

Sugarcane was therefore a unique source of a chemical that gave people intense pleasure when consumed. It could be grown on a large scale and it would give a very high yield of sucrose per acre – sugarcane, like corn but unlike wheat and rice, evolved a mechanism for unusually efficient photosynthesis – but the large amount of labor and energy required for growing the plants, harvesting the stems, and extracting the sugar still made it expensive. For a long time, sugar was treated not as a staple food item but indeed as an expensive spice – something that added flavor – and even as a medicine.[8]

In the 15th century, as the various Christian kingdoms in the Iberian Peninsula of southwest Europe were in the final process of forcibly removing the few remaining Arab kingdoms from the Peninsula, two of these Christian kingdoms, Portugal and Castile, took the war across the Straits of Gibraltar to the African side. The Portuguese in particular decided to extend this official crusade and to encircle the entire Arab dominion by going around Africa to reach India and joining hands with the legendary Oriental Christian kingdom of Prester John to fight the Muslims.[9] One important aspect of this proposed strategy was to wrest the spice trade away from the Arabs, which should greatly weaken them economically. That Venice and Genoa, whose merchants bought the spices from the Arabs and brought them to Europe, would also suffer, was an added bonus to the Portuguese. To achieve this goal, at the beginning of the 15th century the Portuguese embarked on a program of naval exploration with the aim of rounding the African continent, a program that eventually would lead them to India by the end of that century (Chapter 4).

But before they reached India, at the very beginning of this exploration program, the Portuguese ships first arrived at a string of islands along the northwest coast of Africa, including the Madeira Islands and the Canary Islands (Figure 3.2). These two groups of islands had in fact been visited before by the Phoenicians, Greeks, and Romans, but when the Portuguese arrived in the Madeira Islands in 1419 there were no inhabitants there, and only a small population of North African extraction was living in the Canary Islands. The Portuguese soon began to colonize the islands, sending settlers from the Portuguese mainland. They tried grain agriculture on the Madeira group, but this was unsuccessful due to the climatic conditions, so they soon switched to the cultivation of sugarcane, which they had first seen growing in the Arab settlements on the Iberian Peninsula. On the Canary Islands, where the Spanish Castilians contested the Portuguese for possession and eventually won, sugarcane agriculture was also established by settlers, not without a long fight to

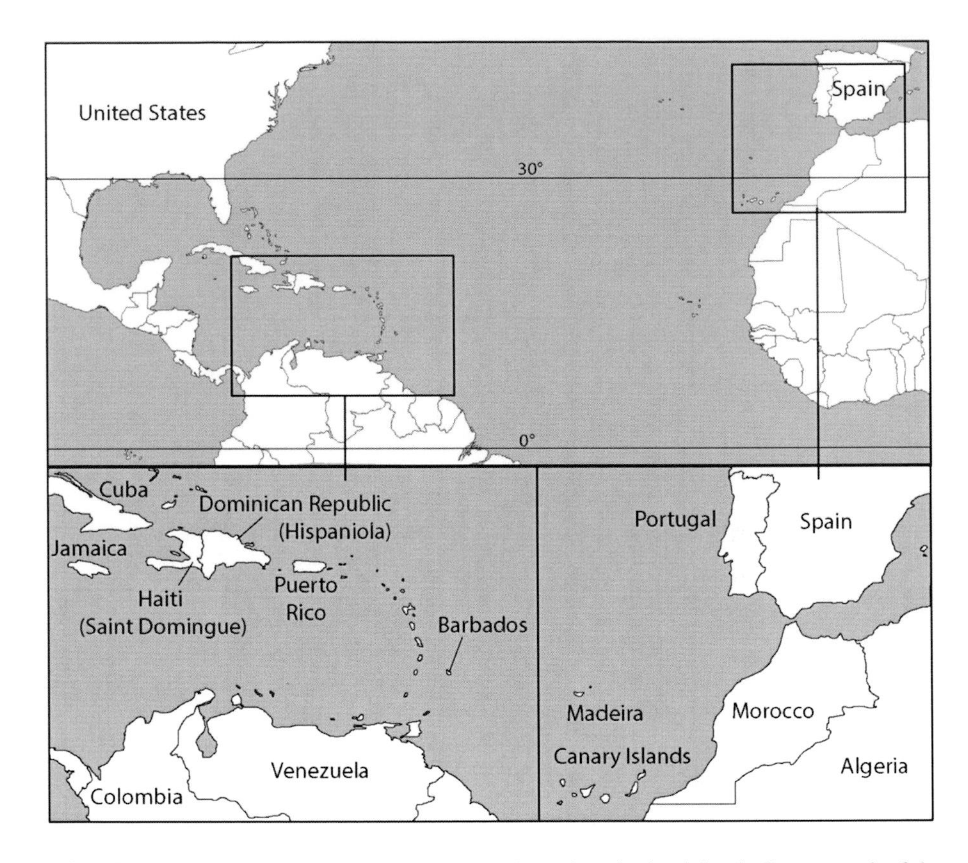

FIGURE 3.2 Locations of colonial sugarcane plantations in the Atlantic Ocean north of the equator (modern state lines are shown).

suppress and dispossess the local inhabitants. Sugarcane agriculture turned out to be well suited for these islands – the two groups of islands are located between the 28°N and 32°N latitude, and they have exactly the temperature and rainfall conditions in which sugarcane thrives. Being outside the Mediterranean, the islands were also away from the reach of the Venetian and Genoese navies, which were jealously guarding their duopolies of the sugar trade with the Arabs.

What these sugarcane colonies needed was labor. The Iberian Christians were familiar with the slavery system that was practiced by Muslims on the Iberian Peninsula and in North Africa, an economic and legal system distinct from the various feudal systems prevailing in most of Europe at the time. The formal approval for enslaving Africans, or anyone who was not a Christian, was given by Pope Nicholas V in 1452. From that point on, the Portuguese, and eventually the Spaniards, began buying large numbers of African slaves from Arab or African middlemen, or directly obtaining them by raiding the nearby African coast. These slaves were put to work in the sugarcane plantations on the Madeira and Canary Islands, as well as in other enterprises on the Iberian Peninsula.

The establishment and operation of sugarcane plantations in these West African islands was the first phase in the development of the system of capitalism in the West.

This phase, which the historian Sven Beckert termed "war capitalism,"[10] although a more comprehensive term would be "war and slavery capitalism," became prominent in the next few centuries and lasted well into the 19th century. War and slavery capitalism was a novel system of agricultural production. Under this system, European states – or their proxies, in the forms of various "companies" – militarily conquered territories outside Europe from the local inhabitants. They then used capital from Europe to establish large-scale plantations that operated as sort of proto-factories with a regimented work force that not only grew the agricultural products but also carried out at least the initial processing steps that made it into a marketable commodity. The capital was also used to buy the slaves that constituted the work force, a crucial aspect of the system that made the price of the product competitive in European markets and elsewhere. While efforts were sometimes made to enslave the local population, these were usually not successful for various reasons (see below, and see also the case of the Dutch spice plantations in the Banda Islands described in Chapter 4), and in most cases slaves were eventually imported from elsewhere, mostly from Africa.

An important aspect of this colonization process, first developed for sugarcane plantations but soon applied in general, was the construction of a legal system, backed up by the military force of the occupying state or its proxies, that treated the European settlers in the same way as if they were living in their original country for handling affairs among themselves, but that gave the natives and the African slaves very few rights vis-a-vis their European masters and the European settlers in general. While such a system had some precedents in the past, most notably with the slave plantations of the Roman Empire and, in a limited fashion, in Arab lands during the Middle Ages, the level of violence and industrialization involved in the sugarcane plantation model was novel. And as it was implemented in the Americas starting at the beginning of the 16th century, it was about to change world history in a drastic, and violent, way.

For a brief time in history, the Madeira and Canary Island groups (and also the Portuguese island of São Tomé on the equator in the Gulf of Guinea, off the coast of West Africa) were the primary suppliers of sugar to Europe, making the plantation owners, merchants, and bankers extremely rich. Then, in 1492, the naval expedition led by Columbus, a Genoese at the service of the Spanish king (the Portuguese turned down his request for sponsorship) discovered the Caribbean Islands, among them Hispaniola, the island which today is divided into Haiti and the Dominican Republic. On his second voyage a year later, Columbus, whose wife was the daughter of a sugarcane plantation owner in Porto Santo in the Madeira Islands, stopped in the Canary Islands on the way west and picked up sugarcane cuttings (called "sets") for a plantation that he planned to establish in Hispaniola.[11]

Columbus had been appointed by the Spanish Crown in advance of his trips as governor of all the lands he would discover, and upon his arrival on that trip he did establish the plantation, whose land was obtained by force from the locals. Columbus delayed attempts to Christianize the locals so that he could use them as slaves (since according to the pope's edict, Christians could not be enslaved), but the locals proved resistant. Harsher measures such as corporal punishment and amputations also did not succeed in compelling them to work to the satisfaction of the

Spaniards. In addition, the local population was quickly diminishing because of overworking, atrocities visited on them by the Spaniard settlers and soldiers (who used the locals for testing the efficacy of their weapons, for example to see if their swords were sharp enough to decapitate a person with one strike), and because of contagious diseases unwittingly introduced by the Europeans and for which the locals had no immunity. The local population on Hispaniola, estimated originally at several hundred thousand, was reduced to a few thousand within a single generation. With the lack of a local work force, African slaves were soon introduced, and sugar from Hispaniola was first shipped to Europe in 1517.

The Caribbean Islands, and other tropical lowlands in the Americas, proved to be ideal locations for growing sugarcane (Figure 3.2). But the Spanish never put a lot of effort into developing their sugar plantations and soon diverted their attention to the mainland countries in Central and South America, where large quantities of precious metals such as gold and silver could be had (with the use of military force and slave labor, of course). Their place in the sugar business was taken first by the Portuguese, who discovered Brazil by chance on their way to India in 1500 (see Chapter 4) and soon afterward began a process of military conquest of the land from the locals, establishing sugarcane plantations with African slaves. Brazil became the major supplier of sugar to Europe during the 16th century.

The Caribbean itself became a place where Spanish, French, English, Dutch, and even Danish navies continuously fought the locals and each other for control of land, which was used mostly to establish sugarcane plantations. In 1627, the first English sugarcane plantations were established on the island of Barbados, and by mid-century these plantations were exporting substantial amounts of sugar to England. When settlers ran out of land to cultivate on Barbados, English forces invaded Jamaica in 1655, took it over from the Spanish, and established numerous sugarcane plantations there too, although battles with the local residents and descendants of slaves freed by the fleeing Spanish colonists, hiding in higher grounds, continued until the middle of the next century.

While the English settlers first tried to employ indentured servants from Europe, they found it difficult to enforce their will on these workers, who would not work as hard as the settlers wanted them to and who often escaped. The settlers soon switched to purchasing African slaves in large quantities. This experience was shared by other European settlers. It is important to note that throughout the history of the plantation system, and not only for sugarcane, there was a very high mortality rate for indentured servants and even more so for slaves. The two main diseases causing death were yellow fever and malaria, both introduced from Africa with the slaves and to which the African slaves, unlike European people, were considered to be somewhat more resistant. However, in combination with overworking and the general mistreatment as well as the tropical climate, these two diseases killed plenty of African slaves. In some colonies such as Saint-Domingue (see next paragraph), 60% of the slaves died within five years of arrival. To this toll must be added the high mortality rate in the passage to the colonies, again particularly high for the African slaves but also not inconsiderable to European indentured servants either.

Besides the Spaniards, the Portuguese (in Brazil only), and the English, the other major participants in this land grab were the Dutch and French. In addition to occupying several Caribbean Islands, the latter two, as well as the English, also carved out colonies on the northeast tip of South America (in present-day Guyana, Suriname, and French Guiana), for the specific purpose of establishing sugarcane plantations, although these mainland colonies were never very prosperous. The Dutch also attacked and conquered the northern part of Brazil from the Portuguese in 1639 but were eventually defeated and had to give the territory back to Portugal in 1654.[12] Overall, the permanent Dutch possessions were relatively small and never amounted to a major sugar industry. The French, on the other hand, had extensive sugar plantations in the Caribbean Islands such as Martinique and Guadeloupe, but their largest and most productive colony was Saint-Domingue, which comprised the western part of Hispaniola, a region of the island that the Spanish more or less vacated on their own. Saint-Domingue (today, the state of Haiti) developed in the 18th century into a huge exporter of sugarcane, as well as coffee, cotton, and some tobacco. On the eve of the French Revolution in 1789, 40% of the sugar consumed in Europe came from Saint-Domingue (and 60% of the coffee). At that time, its population of African slaves was estimated at about 450,000, roughly 90% of the total population. The rest were split evenly between white settlers and free "people of color," mostly descendants of European males and African slave women.

Since the French Revolution extolled the "rights of man," both the white planters and the slaves on Saint-Domingue expected that soon enough the new government in Paris would emancipate the slaves. This expectation led to the unfolding of a complex set of events in the next 15 years. At first, the white plantation owners were in favor of independence from France, so that they could continue the legal system that allowed for slavery. But emancipation was not declared quickly, allying the apprehensions of the plantation owners and disappointing the slaves, who then launched a rebellion in 1791. Emancipation was eventually declared in 1793 in Saint-Domingue by the local representative of the revolutionary government in Paris (apparently on his own initiative), and the French government officially abolished slavery throughout its domain in 1794. But by then, the military conflict in Saint-Domingue had gotten more complicated by the development of civil war and the involvement of Spanish and English forces, who were approached by the white settlers for help in keeping the slaves down. With all this upheaval, sugar production on the island basically ceased, and this meant that France was losing a huge amount of revenue that it usually received from taxes on sugar at the same time that the Revolutionary War in Europe and elsewhere was costing her a fortune. When Napoleon became the ruler of France in 1799, he decided to send a large army of 20,000 soldiers to restore French control and perhaps slavery (as he did on Martinique, although his intentions for Saint-Domingue are still in dispute).

The French army arrived in 1801 and the slaves, under the leadership of Toussaint Louverture and Jean-Jacque Dessalines, initially resisted but eventually signed an agreement with the French general, Leclerc, to lay down their arms. However, after a few months the now ex-slaves realized that the French authorities were intent on restoring slavery, and they resumed their revolt. With the majority of the French soldiers already dead from yellow fever, on November 18, 1803, the slave forces under Dessalines[13] decisively defeated the French army, and a free Republic of Haiti was declared on January 1, 1804. The French, while not immediately recognizing

Haiti as a free country, never again attempted to reconquer it. In fact, Napoleon had already realized, in large part because of the ongoing military conflict in Saint-Domingue, that France did not have the military strength and economic resources to maintain most of its possessions in the New World, and in April 1803 France signed an agreement with the United States to sell it the French possessions in continental North America, a deal known as the Louisiana Purchase. Overall, it is estimated that 200,000 blacks and mixed-race people were killed in the war, as well as at least 100,000 French and English soldiers. While some of the estimated 30,000 white settlers escaped, most were killed either during the war or in massacres conducted by the triumphant slaves after the declaration of independence.

Sugar production in Haiti never recovered, and Haiti ceased to be a major exporter of sugar. The Caribbean, though, continued to be a major center for growing sugarcane in the 19th century. The Spanish possessions in the Caribbean, particularly Cuba and Puerto Rico, had many sugar plantations. England continued to grow sugar on Barbados, Jamaica, and a few other islands. There were also some sugarcane plantations in continental North America. Previously, after the conclusion of the Seven Year War in 1763, England obtained Florida from Spain in exchange for Havana, and English settlers quickly began sugarcane plantations in the territory, which extends up to the 31°N latitude. However, at the conclusion of the American War of Independence, the Treaty of Versailles in 1783 gave Florida back to Spain. But most of the English settlers remained in Florida (as discussed below, Florida Territory was ceded by Spain to the United States in 1821). Also, sugarcane plantations had been established in Louisiana by the French, and they came under American control once the Louisiana Purchase went into effect at the end of 1803. And in all these places, the sugar plantation system continued to be based on capitalism, force, and slavery well into the 19th century.

But the 19th century did witness the gradual abolition of slavery and the introduction of competition from sugar beet cultivation by free labor in temperate zone countries, mostly in Europe, so the sugarcane farming and processing business had to adapt to the new circumstances. Furthermore, with demand for sugar constantly increasing around the world as sugar became a major ingredient in both solid foods and in drinks (such as coffee and tea, Chapter 5), and a major contributor of calories in the human diet, new lands for growing sugarcane were needed. While land grabs of tropical areas by European countries continued, for example in Hawaii (Chapter 7), Natal (part of South Africa today), Fiji, and Australia, in these later imperial extensions slavery was not usually used. Instead, indentured servants by the millions were brought from far away countries, for example from India, Japan, China, The Philippines, and even the Caribbean Islands, setting the stage for future violent conflicts such as the recent civil wars in Fiji between the original inhabitants and the descendants of the Indian sugarcane workers. By the end of the 19th century sugarcane growing also began to mechanize, gradually reducing, but by no means eliminating, the need for manual labor. Today, sugarcane, this highly efficient photosynthesizer, is still an extremely important crop plant. As of 2015, world production of sugar from sugarcane was roughly 1.9 billion metric tons, constituting the most abundant edible farm product on earth, more than double the next product, which is corn. Brazil is still the biggest grower, with 40% of total world production, with India, China, and Thailand also being major sugarcane growers. However, as with corn, much of the sugarcane sugar produced today is converted to ethanol fuel, particularly in Brazil, where 55%–60% of the sugarcane crop goes to ethanol production.

TOBACCO

At the same time that the introduction of sugarcane farming to the tropical parts of the New World was making such an impact on that region, a native New World plant with a much wider growing range was also becoming a world commodity, and in the process making its own distinct impact on the New World and beyond. This was the tobacco plant (Figure 3.3), and its impact on human society was due to a special chemical that it contains called nicotine.

There are actually more than 60 species of plants in the genus *Nicotiana*,[14] most of them found in various locations in the Americas and a few that grow in Australia. These generally herbaceous plants make a chemical called nicotine (Figure 3.4) in their roots and transport some of it to the leaves, stems, and flowers. Nicotine is first and foremost a toxin to animals and microorganisms. It is important to realize that toxicity is a general term, defined simply by the end result. If we apply a certain chemical to a certain living organism and we observe a negative outcome on the well-being of that living organism, we say that the chemical is toxic *to this organism*. In many cases we do not know how the chemical causes the ill effects, but in some

FIGURE 3.3 A tobacco plant.

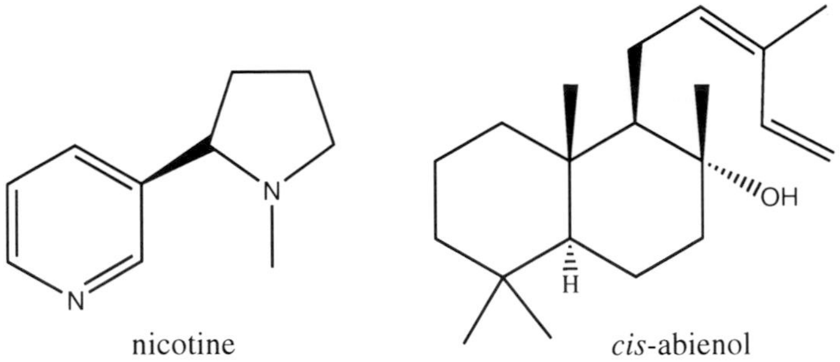

nicotine *cis*-abienol

FIGURE 3.4 The chemical structures of nicotine and *cis*-abienol, both made by tobacco plants.

cases that have been investigated in more detail, particularly in humans but also in other living organisms, a specific mechanism has been elucidated. The amount of the chemical applied relative to the weight of the organism obviously will be correlated with the severity of the effects. For example, a small caterpillar eating the leaves of *Nicotiana tabacum* might die from ingesting a small amount of nicotine, while a person ingesting the same amount of this chemical might suffer no bad effects because the chemical will spread throughout the much larger human body and its toxic effect on human cells, whatever it is, will be highly diluted.

However, it must also be remembered that each living species has a different set of genes, and therefore some species may not be susceptible at all to certain toxins or have evolved genetic and cellular mechanisms that allow them to counteract the effect of specific toxins. Such countermeasures could for example involve the breaking down of the toxin in the liver as humans and other mammals do, or by preventing the absorptions of the toxin into the body in the first place. As it happens, nicotine is quite toxic to people. Its LD_{50} value (LD = Lethal Dose), defined as the amount of the compound per kilogram (kg) of animal body weight at which 50% of the animals die, for humans is generally cited as 1 mg/kg, although it might actually be 10 mg/kg, still an indication of strong toxicity.[15] Indeed, nicotine had frequently been used as a surreptitious murder weapon until a forensic method for its detection in dead bodies was developed in 1850 by Belgian chemist Jean Servais Stas.[16] Furthermore, lethal accidental poisoning from nicotine-containing pesticide solutions was not unheard of prior to the banning of nicotine-containing pesticides in the United States in the middle of the 20th century.

Why do *Nicotiana* plants make nicotine? As was already briefly mentioned in Chapter 2, with additional examples discussed in Chapters 5 and 6, many plants make toxic compounds to protect themselves from pests. Plants are at the bottom of the food chain. Other living organisms, microbes, and animals (of the latter, particularly insects), eat plants for their own survival and growth. Being eaten generally gives plants no advantage (with the exception that having their ripe fruits eaten by birds or mammals often helps disperse their seeds), and in fact it puts them at a disadvantage in the struggle for survival and reproduction. So it is not surprising that plants that have evolved the ability to protect themselves from their enemies have thrived and multiplied, at the expense of plants that have not evolved efficient defenses. The chief defense system that plants have evolved relies on making and accumulating chemicals that have deleterious effects on animals and microorganisms. Once absorbed into the body of the organism that is attacking the plant, either via surface contact or ingestion, these plant chemicals can cause disruption of the normal working of cells or internal organs of the offender, with consequences ranging from minor physical ailments all the way to death. In addition, for insects and other animals, plant defense chemicals could disrupt functions unique to these multicellular organisms, such as those operating in the nervous system or the reproductive systems. Herbivory is what any non-plant living organism does when it eats plants, and while the term has a peaceful, indeed pacifist, connotation in our human languages, as does the synonymous term vegetarianism, the plants see it quite differently. For them, it is war, and they respond with chemical weapons.

Nicotine, being toxic, protects the *Nicotiana* plants from herbivores big and small that try to feed on its leaves, stems, and roots. But in addition to its direct defensive role, another interesting function for nicotine was found in the flowers of *Nicotiana attenutta*, an annual plant that grows in the western United States. The white flowers are visited in the evening by a moth, which finds the flowers by the scent they emit. Once it hovers over the flower, the moth extends its long tongue to the base of the flower, where some "nectar" solution containing sugars is found, secreted from the floral "nectar glands." Because the nectar is found at the base of the flower, and the moth, even with its long tongue, must get very close to the flower, the head of the moth often brushes again the pollen-bearing anthers and as a result pollen grains stick to the moth. After the moth drinks this nectar, it flies away to visit another flower. As it tries to drink the nectar of the new flower, its head, with the pollen grains on it, brushes against the stigma, the female part of a flower, and thereby deposits the pollen grains it picked up from the previous flower it visited. It also picks up new pollen to transfer to the next flower. In this way, plants managed to get insects to help them in their reproduction.

The nectar in the flower is clearly the reward that plants evolved to produce to attract the moths. But it turns out that the plant is able to have a much finer control of the process. It is to the benefit of the plant that each flower is visited by multiple moths that will then spread its pollen to multiple other flowers. So it provides enough nectar for multiple moth visits. However, a moth finding a large quantity of nectar may just stay at this flower and drink as much as it needs – and then just take a nap. But this is not what happens. The moths have been observed to spend just a brief time over each flower and to drink only a limited amount of nectar. Why? Because the nectar also contains a small amount of bitter-tasting nicotine. It has been observed that in flowers that have been genetically manipulated to have no nicotine in their nectar, each visit by a moth lasts much longer and most of the nectar is depleted in a single visit. Thus, the presence of nicotine in the floral nectar appears to have evolved to promote more efficient pollination.[17]

The toxicity of nicotine to microorganisms appears to be the basis for the ancient human practice of using tobacco leaves and juice to dress wounds and drinking the juice to treat intestinal parasites. The reputed efficacy in treating these and other medical conditions by compress and by the internal consumption of tobacco material through smoking, ingestion, and enemas was perhaps one reason for the initial rapid spread of tobacco growing and use throughout the Old World once it was introduced to Europe by Columbus. To be sure, the use of tobacco engendered many negative responses and prohibitions of usage by many authorities, often predicated on the objection to the foul smoke but also on noticing the negative effects of tobacco use on the physical health and mental activity of the users. The latter refers to the clear psychoactive properties of nicotine, which were probably the major cause for the consumption by humans of tobacco in the first place and the reason why this use is still widespread today, despite overwhelming evidence of the serious damage to one's health that long-term consumption of tobacco products entails.

The psychoactive properties of nicotine are numerous and have by now been extensively investigated. In general, at the concentrations of nicotine in the blood stream that are achieved via the usual delivery methods, nicotine acts as a stimulant. One such method is smoking, in which the nicotine is volatilized by the heat of the fire of the burning leaf and upon inhalation binds to the mucous of the cells in the mouth and lungs, thus entering the blood stream. Other common methods include chewing and snuff, in which nicotine makes direct contact with the mucous in the mouth and in the nose, respectively. Nicotine also gives a general sense of well-being, mostly (but not only) by inhibiting unpleasant sensations such as hunger and pain. Nicotine achieves these affects by binding to numerous types of receptors in the nervous system that fall into the general class of nicotine acetylcholine receptors, thus interfering with the transmission of signals to the brain. While people often react negatively to first use of nicotine, the sense of well-being that is experienced upon repeated use is certainly a draw and may be the reason why nicotine is such an extremely addictive chemical. However, at very high concentrations of nicotine the neuroreceptors might be overwhelmed and hallucinations, unconsciousness, and death may follow.

The Amerindians discovered the psychoactive properties of tobacco thousands of years ago,[18] and they used tobacco for religious, social, and recreational purposes. Two particular *Nicotiana* species, both originating in South America, were cultivated, *N. rustica* and *N. tabacum*, and both spread throughout the Americas before Columbus and his men arrived there. It is not clear which species was brought back by Columbus to Europe, but it is likely to have been *N. tabacum*, and this is the species that is today cultivated the most. *N. rustica*, which has about ten times the concentration of nicotine in its leaves as does *N. tabacum*, is also still cultivated but to a much lower extent.

Within a century after the introduction of tobacco plants to the Old World, its cultivation spread everywhere in Southern Europe, the Middle East, Africa, and the Far East. However, the New World continued to be a major center for the production of tobacco, particularly for Spain, England, and the Northern European countries.[19] In the Caribbean basin, Cuba, a Spanish colony, became a center of tobacco growing. The Caribbean Islands, as described above, lost most of the local inhabitants soon after the Europeans arrived, and tobacco cultivation, like sugarcane cultivation, was only possible by massive importation of African slaves. The presence of both sugar plantations and tobacco plantations in Cuba made it a coveted target for other imperial forces in the area. Certainly, when the English briefly occupied Havana in 1762 before giving it back to Spain the next year, they quickly began to purchase and export as much tobacco as they could out of Cuba. And the tobacco allure was probably a factor in the American motivation for the aggressive military moves that led to the occupation of Cuba at the end of the 19th century.

But the place where tobacco was at the center of multiple armed conflicts was in North America. The first successful English settlement in North America, Jamestown, began in 1607 in what later became the colony of Virginia. Initially established as a trading post to obtain gold, it soon became clear that no gold was to be had in the area, and if the colony were to survive in the long run it would have to develop an economy based on agriculture. The settlers soon hit on the idea of growing tobacco. They first tried *N. rustica*, but this tobacco has a strong and "harsh" flavor that was disliked by most European smokers. One of the settlers, John Rolfe (who later married Pocahontas, the daughter of the local Amerindian chief Powhatan), was able to obtain *N. tabacum* seeds from the island of Trinidad – overcoming a Spanish ban on the export of such seeds – and by 1612 the Jamestown colonists were exporting tobacco leaves to England.

Cuban tobacco and North American tobacco – generally referred to as Virginia tobacco – were highly desired products believed to have superior flavor to tobacco grown in other parts of the world, and to some extent they still keep this reputation. In the wine industry, *terroir* is a term that connotes a particular constellation of soil and climate that determines the taste of the wine, a property that appears quite subjective and impossible to quantify. The concept of *terroir* has also been applied to tobacco, to explain the reputations for high quality of Cuban and Virginia tobacco, reputations that may suffer from the same level of subjectivity as the judgment of wine flavor.

Be that as it may, it is important to realize that tobacco flavor, as separate from the psychoactive effects of nicotine but perhaps contributing to them, depends to a large extent on the many other compounds that are found in the tobacco leaf in addition to nicotine, many of them defensive/toxic compounds in their own right. Because the nicotine molecule has nitrogen atoms in it, and nitrogen, in a form that the plant can use, is energetically expensive for the plant to obtain and is often simply not available in sufficient amounts in the soil, the plant limits the amount of nicotine that is usually found in the leaf. Instead, it makes nicotine in the roots and stores most of it there, transporting only a bit to the stems, leaves, and flowers unless the plant is attacked by herbivores or pests. Once the plant is attacked, more nicotine is mobilized from the root to the aerial parts of the plant. But until it encounters the pests, the plant saves the nicotine, and instead relies heavily on several other toxic compounds that are more economical for the plant to make. These include a compound called *cis*-abienol (Figure 3.4), which, like nicotine, also volatilizes and gives the tobacco smoke a distinct smell that many people find very pleasant, as well as many other compounds that affect the smell of tobacco. Furthermore, tobacco plants have a genetic system to detect attack by herbivores, and when such an attack occurs, the plant responds by synthesizing even more toxic compounds, many of which can also volatilize.

Thus, the overall "flavor" of tobacco is determine by the synthesis of many compounds, controlled by the information found in the genes of the plant, the soil, the climate, and the specific local pests, and, most importantly, the curing process that's employed.[20] It is also important to remember that tobacco "flavor" can be experienced not only via smoking but also by the various methods of direct application of the leaf, such as chewing and snorting (snuff), although the various methods usually do not give the same sensation. It may not be fully appreciated today, but until the beginning of the 20th century less than half of the tobacco material sold in the United States was smoked, and cigarettes, the most common tobacco product today, did not become very popular until World War I.

Once the Virginia colony settled on agriculture for its main source of income (rather than trade as originally intended), the settlers were bound to run into conflict with the local Amerindian population over the use of the land. However, the choice of tobacco as the main agricultural crop caused the conflict to happen sooner rather than later. The local population, which was probably already thinned out because of contagious diseases unwittingly introduced by the Europeans, used the mostly fire-managed forested area for some basic agriculture and for hunting. The settlers, on the other hand, wanted to clear more and more forest to grow their tobacco. In addition, tobacco plants exhaust the nitrogen in the soil – always a limited resource for plants – faster than the average crop, because nitrogen is a major component of nicotine. Nitrogen can be added to the soil in the form of fertilizer such as animal manure, but this is also relatively expensive (even today, synthetically made nitrogen fertilizer is a major expense for farmers). Certain crops such as the legumes (including peas and alfalfa) actually add nitrogen to the soil, so a system of crop rotation, in which tobacco crop is grown in a given field one year and then another crop such as a legume is grown the next year, is also a viable solution. But tobacco was bringing the settlers a lot of money, so they employed a different solution, which was simply to cut down the forest and grow the next year's tobacco crop on the newly cleared fields. The move to new land had another advantage – escaping many tobacco pests that got established in the old tobacco fields.

The Amerindians realized quickly that the settlers were set on an inexorable expansion, and they began to resist. In 1622, in a coordinated attack on Jamestown and the smaller surrounding settlements, they managed to kill about 400 settlers, about a quarter of all settlers in the colony at the time. The Amerindians expected the settlers to conclude that they would not be able to hold the settlement and therefore should abandon it, but this did not happen. Instead, after some additional minor skirmishes a truce was declared. The settlers invited the Amerindians to celebrate the peace agreement in Jamestown and served them wine laced with poison, killing 200 of them this way and finishing off 50 more with their weapons. After this, major hostilities ceased for about 20 years, but in 1644 another 500 settlers – about 10% of the settler population at the time – were killed in a string of attacks by Amerindian warriors. Again, the colonists retaliated, killing many Amerindians and capturing and killing their chief.

The pattern was set for the next 200 years. By the 1630s, the land grab for tobacco cultivation by English settlers spread northward to what became the Maryland colony, southward into the future Carolina colony, and of course westward. During that time, there were numerous military clashes with the local Amerindians who naturally objected to being removed from their land. Often, a local treaty would be signed, only to be violated by individual settlers who pushed into Amerindian territories, followed by violent attacks on these settlers by Amerindians. Once settlers were attacked, a military reprisal by the armed forces of the settlers, often culminating in the forced removal of the Amerindians from the area, was the most common course of action. In this way, the Amerindian population was soon entirely removed from the coastal areas, and in their place large tobacco plantations with many slaves and indentured servants were established. These plantations were particularly large and successful in the Chesapeake Bay area and in other coastal river basins, since for a long time water transportation was the most economic method of shipping tobacco to market.

Because the price of tobacco was relatively low, large plantations where economy of scale could be achieved had an advantage. However, unlike sugarcane cultivation, the start-up costs for growing tobacco were small and neither was a major capital outlay needed for equipment to process the tobacco leaves. Seeds were plentiful and cheap – the tobacco seeds are tiny, and a single plant allowed to flower and set seeds produces enough seeds to plant several acres. Furthermore, while tobacco cultivation was labor-intensive, unlike sugarcane cultivation (or cotton cultivation, as we'll see next), most forms of tobacco cultivation and processing did not require a particularly large workforce at specific times of the year, to perform a few critical tasks that must be finished quickly. Rather, the need for labor was more or less spread throughout the year, starting with germinating seeds in a seedling bed, planting the seedlings in the field, tending to the plants as they grew by trimming, weeding, and removing pests, and then harvesting the leaves and curing (i.e., drying) them in simple storage facilities before packing them and transporting them for sale. This meant that a farmer with small means, with his family or just a few slaves or indentured servants, could run an economically viable tobacco farm, particularly if land could be had for very little. These specific conditions of tobacco farming gave impetus to the continuous westward push of the European settlers.

According to the original royal charters of the Virginia and Carolina colonies, their borders on the west were not clearly defined. As the westward expansion of the tobacco settlers in the Virginia Colony and the Carolinas (the Carolina colony officially split into North Carolina and South Carolina in 1729) was causing friction with the Amerindians, the British Crown endeavored to address this problem with the Royal Proclamation of 1763. Given in conjunction with the conclusion of the world-encompassing Seven Year War (a war that to the English colonists in North America was also known as the French and Indian War) that set the Mississippi River as the border between the English possessions in the east and the French possessions in the west, it forbade English settlements west of a line that ran roughly north–south along the Appalachian Mountains (Figure 3.5). While this edict was violated by individual settlers as soon as it was made public, once the independence of the colonies was established it was completely ignored. Thus, there were soon so many settlers in the western parts of Virginia and North Carolina, many of them tobacco farmers using the Mississippi River as the main route to transport their product, that these areas were made respectively into the state of Kentucky in 1792 and the state of Tennessee in 1796. In both of these new states, clashes with local Amerindians continued, leading to thousands of casualties on both sides. Eventually – during the first half of the 19th century – a series of large-scale wars and expulsions of the Amerindian

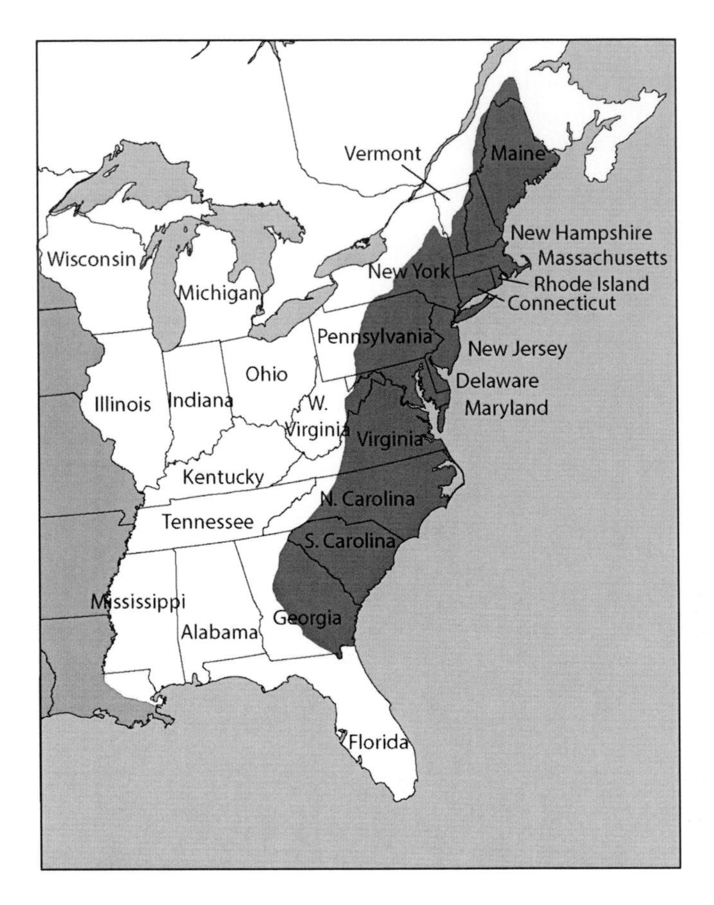

FIGURE 3.5 A portion of eastern North America superimposed on today's state borders, showing the 1763 "Proclamation Line." This line is the border between the dark-shaded area, where the English colonies were allowed to spread, and the white area, controlled by England but reserved for the endogenous Amerindian population. The light-gray area on the left was under French control.

communities occurred in these areas, in the parts of South Carolina that became Georgia, Alabama, and Mississippi, and in Florida. But since these later events were tied more closely to another major plant crop whose cultivation was quickly expanding by the European settlers – cotton – they are discussed later in this chapter.

But before we move on to cotton, a 20th-century armed conflict within the United States involving tobacco growing is worth recounting, even though the name given to it, the Black Patch Tobacco War, is a bit exaggerated. The origin of the conflict lay in part with the invention of the modern cigarette and, starting from the 1850s, the widespread adoption of this smoking device first by soldiers but eventually by many other members of society. At first, cigarettes, which are made from shredded tobacco leaves wrapped in paper, were prepared by hand in a laborious process that kept cigarette prices relatively high. Then in 1881, American James Bonsack invented a cigarette-rolling machine that sped up the process a hundredfold.

This made cigarettes very cheap. In 1885, American James Duke, who owned a cigarette-making factory, obtained exclusive license to use Bonsack's machine. As a result, he was able to undersell all his competing tobacco manufacturing companies in the United States. Using other anti-competitive business approaches, such as controlling the tobacco-flavoring manufacturer of licorice and tobacco type branding, Duke was able to both bring most of the American tobacco manufacturers under his control, establishing the American Tobacco Company (ATC), and segmenting the market for the tobacco produced by the farmers. ATC then used its monopoly to force farmers to sell it their different tobacco types at extremely low prices, financially ruining many farmers.

In 1904, in a region encompassing several counties in Kentucky and Tennessee called the Black Patch (because the method used to cure the tobacco growing there was called "dark-curing"), a farmers' cooperative called the Planters' Protective Association (PPA) organized to fight against the ATC monopoly. The cooperative was hoping that by their unified refusal to sell their tobacco to ATC at the prices demanded by ATC, the company would be forced to raise the price it paid. To increase their chances of success, they tried first by peaceful means to convince all farmers in the area to join the cooperative and to refuse to sell tobacco to ATC at prices set by ATC. However, ATC would not budge, and eventually the members of the cooperative resorted to violence, forming a paramilitary unit called the Silent Brigade, also known as the Night Riders. Between 1905 and 1910, units of the Silent Brigade raided farms belonging to farmers that sold tobacco to ATC and the storage facilities of the ATC company and its affiliates, destroying crops, farm equipment, and animals, and torching large amounts of stored tobacco. Although these operations rarely resulted in the loss of human life, they caused a substantial amount of damage to property. In 1908, the Kentucky governor dispatched the Kentucky National Guard to raid and arrest many of the Night Riders and to guard possible targets. These actions led to a great decrease in the number of violent raids. In 1911 the Supreme Court found that the ATC was indeed a monopoly and decreed the breakdown of the company, and tobacco prices began to rise, basically ending the violent part of the PPA struggle.

COTTON

In addition to fiber from animal hair (wool) and silkworms (silk), fiber from plants has been a major source for making thread for weaving, a technology that evolved independently in both the Old and New Worlds. Unlike wool and silk, both of which are polymers of amino acids (i.e., these fibers are proteins), plant fiber is a polymer of glucose called cellulose (see Chapter 2). Plant cells synthesize cellulose by linking thousands of glucose molecules end to end. Many linear cellulose polymers are simultaneously synthesized in parallel, and these parallel polymers have the tendency to stick to one another, thus forming a fiber. These fibers are deposited on the surface of living plant cells, forming, together with some other chemicals (for example, lignin, which we will discuss in Chapter 6), a hard "cell wall" that physically protects the plant cell from assault. The cell walls of adjacent plant cells are often fused together, creating long cellulose polymers. Humans have used such polymers to make thread for thousands of years, for example by obtaining long fibers from the stems of the flax plant.

A special example of easily accessible cellulose fibers is found among plants in the genus *Gossypium*, the cotton genus. While these plants make cellulose to cover all their cells anywhere in their body, as all plants do, cotton seeds make particularly good fibers of almost pure cellulose that protrude out and away from the seeds.[21] Cotton plant flowers have four or five complex ovaries, also called carpels, with each carpel containing between 8 to 12 ovules, each containing an egg cell. Once the egg cells are fertilized, the embryos begin to develop, and many cells, called trichomes, begin to grow outward on the surface of each ovule structure, which by now has become the seed coat. The trichome cells grow up to 2–5 cm in length without dividing (which is much longer than the length of an average plant cell), all the while synthesizing and depositing cellulose on their surface. As the seed matures, the cells encompassing the ovule tissue, including the trichomes, die and dry up, leaving the seed covered with a fluffy coat of cellulose fibers (Figure 3.6).

What is the advantage to the plants for making these cellulose fibers? The seeds of many plants are in fact "hairy." It is believed that the function of these hairs is to help in the dispersal of the seeds. For example, long fibers attached to one end of the seed and arranged in somewhat conical fashion, as in the case of the seeds of the common milkweed plant, *Asclepias syriaca* (Figure 3.7), help the seed become airborne and travel by wind. A particularly dense mass of fibers makes some seeds, including cotton seeds, "sticky," and they will cling to the fur of passing animals or the feathers of

FIGURE 3.6 Branches of a cotton plant, showing flowers and fruits at different stages of ripening.

FIGURE 3.7 Fibers are common attachments to plant seeds, helping in their dispersal. Here, a seedpod and partially released seeds of the milkweed plant are shown.

bird and thereby "hitch a ride" to somewhere else. The ball of fibers that cotton seeds are embedded in also gives them buoyancy in water. It is indeed believed that from their origin in Africa, cotton plants have spread far and wide by bird dispersal and by floating on the oceans, reaching not only Asia but also the Caribbean Islands, the American continents, and eventually even islands in the Pacific Ocean.[22]

While cotton seeds are not unique in having cellulose fibers attached to them, their fibers do appear to be the best kind for making thread for weaving, as evidenced by the fact that plants in the *Gossypium* genus independently became the main source for thread in three geographically distinct areas of the world in the last several thousand years – in Africa, in India, and, possibly more than once, in the New World. In India, the species *G. arboreum*, whose cotton fibers average about 2.5 cm, was first cultivated 5,000 years ago and from there its cultivation spread to the Near East and eventually to southern China as well. In Africa, *G. herbaceum*, whose fibers are a bit shorter, was grown. In the New World, *G. hirsutum* (upland cotton), whose fiber length is just over 2.5 cm, originated in Mexico and its use spread south all the way to Peru, north to North America, and east to the Caribbean Islands. *G. barbadense* was domesticated in the Caribbean Islands and the length of its fibers reaches up to 4–5 cm, particularly in the varieties known as Sea Island, American Pima, or Egyptian cotton (the latter is a misleading name, derived from where it was introduced to in the 18th century).

Regardless of the species, the different human cultures that independently domesticated cotton also independently developed remarkably similar techniques to make cotton thread. Once the cotton fruit, also known as the "boll," dried up and split open, revealing the four to five individual balls of fibers (each corresponding to a carpel, and containing the seeds inside), the bolls were picked by hand. Next, force was used to tear the cotton fiber (known as "staple") from the seeds, the fibers were pressed together ("carded"), and then simultaneously twisted and pulled ("spun") to form a thread. While each fiber is only a few centimeters long, by staggering the fibers (in the carding process) and then twisting them together (each fiber is naturally twisted to begin with), the thread can resist shearing, helped by the molecular attraction force between parallel cellulose polymers that is due to a chemical principle called hydrogen bonding.

Two thousand years ago cotton cultivation and weaving were well established in Africa, Asia, and the Americas, while in Europe, an area where cotton plants cannot grow, flax was the main fiber for weaving.[23] Cotton thread and cloths have many attributes that make them attractive to people and often superior to flax thread, among them the ability to absorb water and also the ease with which they can be dyed. But from a purely economic point, perhaps the major advantage of cotton, which it shares with flax, is that it is cheaper to produce than wool or silk, two fibers that are made by animals that must be fed with plants.

Cotton plants need warm climate and plenty of water to grow. They can grow further north than sugarcane – they do well up to the 37°N latitude – but they need warm soil (10°C –15°C, or 50°C –60°C) to germinate and 200 frost-free days to reach the point when the cotton can be harvested. While Herodotus in his *Histories* (5th century BC) mentioned Indian cotton, ancient Europeans rarely encountered cotton cloths, which were expensive items obtained through trade. The Arabs brought cotton cultivation to Spain and Sicily in the 8th century, and some cotton spinning and weaving was introduced into northern Italy and beyond as a consequence of the Norman conquest of Sicily in the 12th century, where the Normans encountered this technology. Cotton spinning and weaving technology eventually arrived in England in the early 17th century, but India remained both a center of cotton cultivation and cotton cloth manufacturing until late in the 18th century. Indeed, in India, as in most other countries where cotton plants grew, traditionally the entire process from seed to cloth was a local affair engaged by the farmers and their immediate family – with rural women spinning the cotton grown on their farm and also doing the lion's share of weaving. Indian cotton fabric and finished cloths came in dizzying varieties and qualities and were some of the most desired trade items brought to Europe by Arab traders.

The arrival of European ships in Indian ports starting in the late 15th century increased the importation of cotton clothing to Europe. The African slave trade was a big impetus, as Indian cloths were highly desired items of trade in Africa in exchange for slaves (who were shipped to the Americas to work in the sugarcane plantations, and later in the tobacco and cotton plantations too). The British East India Company (EIC), originally established in 1600 to partake in the Far East spice trade (Chapter 4), eventually settled on the Indian subcontinent as a major center of its commercial activity, and cotton cloths and fabrics were a main focus. Initially the company "factors" (agents) tried to control the cotton market in India, militarily monopolizing certain areas and forcing producers to sell to the company at prices determined by the company. They also tried to move spinning and weaving activities to the cities and into workshops, the better to control the workers and the market and thus fix the prices to their advantage. However, these efforts were not particularly successful for various military and social reasons and the Indian cotton farmers and manufacturers were able to maintain some independence.

EIC, as well as other traders, did bring some raw cotton into England, allowing for a local cotton spinning and weaving industry to develop. In 1721, the English cotton industry was helped by a law prohibiting the sale of some popular fabrics and cloths imported from India. However, by late 18th century, as the English cotton industry grew and cotton cloths became cheap and popular, particularly following the mechanization of cotton spinning and weaving (first powered by water, and then by steam), a shortage of raw cotton developed in England. There was a need to grow more cotton for the English mills but also for cotton mills operating in other European countries. In consequence, cotton cultivation indeed spread around the world and also increased in countries already growing it such as India. But the region of the world that saw the greatest increase in cotton production in the first half of the 19th century, and from which the English mills in particular received most of their raw cotton, was southeastern North America.

The first European settlers in the New World quickly adopted cotton cultivation from the local inhabitants, first growing it on their plantations for domestic use but soon exporting some to Europe, a trade that grew as England, and other European countries, needed more and more raw cotton. By 1760 cotton from the New World constituted about 50% of the raw cotton imported by England, but to the European settlers it was still not a major crop like sugarcane in the Caribbean Islands or tobacco in the North American colonies. But when the demand for raw cotton in England began to grow exponentially in late 18th century, the settlers saw their chance. Land for cultivation of cotton – rich soil situated near rivers for irrigation and easy transportation – was plentiful in the areas comprising the southeastern colonies (and later American states) in the North American mainland, mostly in what were then Georgia and the Carolinas. The economic model of large plantations with many slaves was also well established by then, and the invention of the cotton gin in 1794 to separate the fibers from the seeds increased productivity of cotton farming by many folds, making it much more profitable. The sole difficulty, as far as the settlers were concerned, was that the land was contested on several levels, first among the various European countries that claimed them and, of course, the Amerindians who lived on these lands.

The land grab by the European settlers for the purpose of growing cotton proceeded in a similar way to the previous one involving tobacco but on a much larger scale. As with tobacco farming, it mostly proceeded by settlers continuously spreading west and establishing strongholds on the ground, with the political establishment, the local militia and the country's army following them and cementing control. It went on until 1845, when the state of Texas, which gained its independence a few years earlier in a bloody war between the formally "independent" American settlers and the army of the now independent State of Mexico, was admitted into the Union.[24]

The first major military conflict with the local Amerindians occurred in 1813–1814 and was named the Creek War. By that time, the territories of Alabama and Mississippi were already carved out of the old Georgia and South Carolina colonies, although the southern borders of Georgia, Alabama, and Mississippi were still contested by the Spanish. The Creek (Muscogee) Indians, a federation of tribes, initially were widespread in the area. However, by 1805, following a string of forced treaties in which they gave up much of their territory, these Amerindian tribes were restricted to parts of Alabama and Georgia, and pushed farther south into Spanish territory. As a result of the continuous westward expansion of the American settlers, tensions within the Creeks arose, with some wanting to militarily resist the settlers and some opposing military resistance as futile and counterproductive. The cause of peace was not helped by the fact that the party of military resistance among the Creeks, called the Red Stick party, received some encouragement and logistic support from Spanish and English forces, the latter involved in their own war with the Americans at the time.

Initially, military clashes occurred mostly between the two Creek factions, but white settlers were sometimes attacked, giving the various state militias an excuse to intervene. Andrew Jackson, first commanding troops of the Tennessee militia and eventually Federal troops, played a decisive role in the war. Jackson's American troops, with help from their allies among Creek and Cherokee Indians, defeated the Red Stick Creeks in a series of battles culminating in the Battle of Horseshoe Bend on March 27, 1814. Following this victory, Jackson forced the Creeks to sign the Treaty of Fort Jackson in which they gave up a huge territory, including areas inhabited by Indian groups that fought alongside him against the Red Sticks (Figure 3.8).

After the conclusion of the Creek War, Jackson continued south to fight the Spanish, who were contesting the validity of the Louisiana Purchase, and removed them from present-day coastal Alabama, a war that was relatively bloodless. However, American forces, some under Andrew Jackson, continued to invade Florida between 1814–1819 on the pretext of battling escaping Creek Indians and African slaves, a series of events known as the First Seminole War. The Spanish ultimately ceded Florida to the United States in 1821.

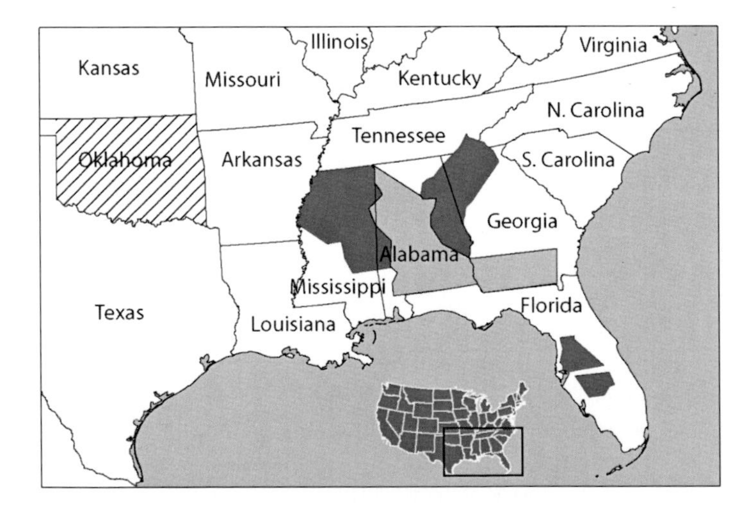

FIGURE 3.8 Dispossessing the Amerindians from the southeast of North America. The area ceded by Amerindian tribes in the Fort Jackson treaty of 1814 is shown in light gray. The Amerindians were expelled from the areas shown in dark gray during 1830–1835, following the Indian Removal Act of 1830. They were sent to Indian Territory (hatched area), which is today the state of Oklahoma. Present state lines are shown in this map.

The Creek War was only a step in the near total removal of the Amerindian inhabitants of the South to make room for cotton farming. Settlers were hungry for land, and in 1830 Congress passed the Indian Removal Act with the full support of Andrew Jackson, who was by then the president. The law allowed the states, using their militias if necessary, to remove Amerindian tribes from their territories, regardless of previous treaties, and to resettle them in an area west of the Mississippi dubbed "Indian Territory" (present-day Oklahoma). Between 1831 and 1838, the Choctaw, Creek, Seminole, and Cherokee tribes were forced to give up their lands in the southeastern United States (Figure 3.8) and to walk all the way to Indian Territory, with many of them succumbing on the way from hunger, exhaustion, and disease (hence the term "Trail of Tears" that was applied to these series of marches). While not all the specific expulsions were directly related to cotton – for example, the expulsion of the Cherokee Nation in 1838 was precipitated by the discovery of gold in their Georgia land – the end result was that huge amounts of land became available for cotton cultivation.

By 1860, cotton plantations were spread throughout the southern United States from the Carolinas, Georgia, Alabama, and Mississippi to Texas, Louisiana, and Arkansas, forming the Cotton Belt – between the Tobacco Belt to the north and the Sugar Belt to the south – and spreading slavery with it. In that same year, total cotton exports from the United States were valued at $192 million, constituting 60% of all American exports. By comparison, tobacco exports were valued at $16 million, and wheat export at $4 million. And then the American Civil War came. While the causes of the Civil War were many, cotton farming and its strong reliance on slavery in the South clearly played a major role in creating and exacerbating the economic and cultural differences between the industrialized North and the agrarian South. The heavy reliance of the South on income from cotton also doomed their Confederacy. The North immediately blockaded the South, and very little cotton was shipped from the South to English or other European ports during the war, devastating the Confederate economy and ensuring its defeat.

As mentioned earlier, even before the Civil War in the United States, the cultivation of cotton, a very lucrative cash crop, was spreading around the world. During and after the American Civil War, cotton became a major crop in the Middle East, particularly in Egypt and other parts of the Turkish Empire along the Mediterranean coast, in Australia, and cultivation increased in India as well. In some cases, producing countries developed a local cotton industry,[25] but many countries simply shipped all the raw cotton they produced to Europe to fill the void caused by the American Civil War. Eventually, cotton production in the United States also recovered, but it never regained its commanding world market share. And the lure of land suitable for growing cotton continued to tempt militarily strong countries, contributing to decisions to engage in imperialistic and colonial ventures around the world. These ventured included the expansion of the Russian Empire during the 19th century into south central Asia (present-day Kazakhstan, Uzbekistan, and Turkmenistan),[26] German occupation of Togo in Africa in 1884, and the Japanese occupation of Korea in the early 20th century.

One aspect of cotton that has directly contributed to warfare is worth mentioning here. In 1846, the Swiss chemist Christian Friedrich Schönbein developed a process to make what eventually became a smokeless gun powder by treating cotton with nitric acid, a process that results in nitrocellulose, also called guncotton. The original gun powder, or black powder, was invented in China around the 9th century, and it consists of sulfur, charcoal (produced by controlled burning of wood, see Chapter 8), and potassium nitrate (saltpeter). Gun powder was used extensively in guns and other weapons until the end of the 19th century, but it has some serious drawbacks; in particular, it leaves too much residue that can clog the gun barrel and interfere with the moving parts of the weapon, leading to jamming and the occasional explosion of the gun barrel. Since the beginning of the 20th century, cordite – the general term given to the combination of nitrocellulose with one or two other ingredients (such as nitroglycerine, nitroguanidine, and petroleum jelly) – has replaced gun powder as the main propellant in ballistic weapons.

CONCLUSION

Sugarcane, cotton, and tobacco are three plants that uniquely, or at least most efficiently, satisfy three different human needs. These needs are in fact complex and perhaps difficult to define. Clothes obviously have the function of protecting people from the elements, but they also play important social functions. Nicotine and sucrose impart specific neurological effects on humans. Interestingly, tobacco in particular has had an important direct effect on wars, as nicotine is a drug whose use is widespread among soldiers and its consumption is often actively supported by the military authorities, since it makes soldiers more tolerant to stress and fatigue.[27] Sugar can also play an important role in nutrition as a source of calories, and as we will see in the Chapter 5, together with tea, which contains another psychoactive drug, caffeine, it helped cotton mill owners to overwork their underage and undernourished workers. But besides the direct effects of these plant materials on individual members of society, since their widespread use began they have had profound impact on the character and direction of the development of human societies around the world. Can we even begin to imagine our society without sugar? Books, movies, and paintings without any smokers? Nobody dying from lung cancer or obesity? No cotton clothes?

Whatever the reasons for the desirability to consumers of the products made from these three plants, their cultivation and trade under the economic systems described here presented opportunities for making huge monetary profits, albeit at the price of enormous losses in human life and the infliction of immense misery on an untold number of people. As is often seen in history, the people who benefitted the most came from the higher social strata of the stronger nations involved. The plantation system that was implemented to grow these three crops – based on the theft of the land from the local inhabitants and the theft of the labor (and freedom) of the slaves and backed up by the force of the state or the state's proxies – provided huge profits to owners, bankers and merchants. Again, from a biologist's point of view, money is a proxy for evolutionary success, as these resources allowed their owners to have more progeny and to care for them well. Perhaps it is also no biological coincidence that many planters – although obviously lacking any understanding of evolutionary biology and perhaps even any self-awareness – sired many progeny with their most often unwilling female slaves (an action which even many contemporaries viewed as an enormous moral wrong, regardless of whether actual violence was involved).

While these are pretty strong reasons why the individuals benefitting from this evil system liked it, why did states support and uphold this mercantilist system – not only with force, but also with tariffs on foreign competitors – for so long? One reason surely had to be that the people personally benefitting from the plantation system were often members of the same small class of people in control of each state in the first place. For example, the Virginians Washington, Jefferson, Madison and Monroe all had tobacco plantations and owned slaves, so it is not surprising that slavery was allowed in the American constitution that they helped write. But more importantly, these three crops typically constituted such a large part of the economy of the countries involved that the taxes levied on these goods became essential to the economical survivals of both exporting and importing countries. Some of these economic factors are still at play today in many countries.

NOTES

1. Levetin, E. and McMahon, K. 2012. *Plants and Society*, 6th ed. McGraw-Hill.
2. The biochemical reaction that links glucose to fructose to give sucrose is a bit complicated, using an "activated" version of the glucose rather than simply glucose. See Heldt, H. W. and Piechulla, B. 2011. *Plant Biochemistry*, 4th ed. Elsevier.
3. Summarized in Diamond, J. 1997. *Guns, Germs, and Steel*. W. W. Norton & Company.
4. A general description of sugarcane cultivation can be found in Mintz, S. W. 1985. *Sweetness and Power*. Penguin.
5. Interestingly, some carnivores, and most birds, do not have sweet receptors and apparently cannot taste sugars, due to the loss of the ancestral gene encoding the sweet receptor. However, hummingbirds have evolved a new receptor for sugars (see Baldwin, M. W., et al. 2014). Evolution of sweet taste perception in hummingbirds by transformation of the ancestral umami receptor. *Science* 345: 929–933.
6. Inborn preference for sweetness is well established (Mennella, J. A. and Bobowski, N. K. 2015. The sweetness and bitterness of childhood: insights from basic research on taste preference. *Physiology & Behavior* 152: 502–507), but whether attraction-avoidance preference is entirely inborn, entirely culturally acquired, or a mixture of both, for the other four tastes is still an open question.
7. And in the 19th century, sugar beet. In the 18th century it was discovered in Germany that the taproot of the sugar beet plant had a concentration of 1.5% sucrose by dry weight. Napoleon began a research and breeding program on sugar beet because of the English blockade that prevented sugar import to France from Frances's Caribbean plantations, and varieties with 5%–6% sugar concentrations were soon developed by plant breeders. Today, sucrose concentrations in some sugar beet varieties exceed 20% of dry weight. Sugar beet is a temperate zone crop and its cultivation is highly mechanized.
8. For a long time, the general distinction between spices and medicines was not very clear. The definition of "medicines" relates to the complex issue of the definition of diseases, and the definition of cure, both of which are constantly changing.
9. Crowley, R. 2015. *Conquerors: How Portugal Forged the First Global Empire*. Random House. In the 15th century the Portuguese were not aware that Islam had spread east from the areas were Arab lived, so "Arab" and "Muslim" were interchangeable.

As described in detail in Chapter 4, on arriving in India the Portuguese met Muslim kingdoms of non-Arabs, and later encountered Muslim principalities all the way to the Spice Islands in what is today Indonesia.
10. Beckert, S. 2015. *Empire of Cotton*. Alfred Knopf.
11. Mann, C. C. 1493. Vintage Books.
12. This sequence of events is sometimes referred to as "The Sugar War."
13. In 1802, Louverture was betrayed by his people and was arrested by the French. He was sent to France and died there in jail from an undetermined cause.
14. The genus *Nicotiana* is in the family Solanaceae, which includes species such as tomato, pepper, potato, petunia, and eggplant.
15. Mayer, B. 2014. How much nicotine kills a human? Tracing back the generally accepted lethal dose to dubious self-experiments in the nineteenth century. *Archives of Toxicology* 88: 5–7.
16. Wennig, R. 2009. Back to the roots of modern analytical toxicology: Jean Servais Stas and the Bocarme murder case. *Drug Testing and Analysis* 1: 153–155.
17. Kessler, D., Gase, K., and Baldwin, I. T. 2008. Field experiments with transformed plants reveal the sense of floral scents. *Science* 321: 1200–1202.
18. Le Couteur, P. and Burreson, J. 2003. *Napoleon's Buttons*. Jeremy P. Tarcher/Penguin.
19. General history of tobacco cultivation and commercialization, with emphasis on various aspects, can be found in Cosner, C. 2011. *The Golden Leaf*. Vanderbilt University Press; and Hahn, B. 2011. *Making Tobacco Bright*. John Hopkins University Press.
20. Nowadays tobacco companies may add to their tobacco up to 500 other "flavor" and "flavor-enhancing" compounds not naturally found in tobacco. Many of these additives impart or enhance scent. Nicotine itself, however, is not usually detected as a smell by the human nose.
21. Tiwari S. C. and Wilkins T. A. 1995. Cotton (*Gossypium hirsutum*) seed trichomes expand via diffuse growing mechanism. *Canadian Journal of Botany* 73: 746–757.
22. See Stephens, S. G. 1966. The potentiality for long range oceanic dispersal of cotton seeds. *American Naturalist* 100: 199–209; and Weigier, A., et al. 2011. Recent long-distance transgene flow into wild populations conforms to historical patterns of gene flow in cotton (*Gossypium hirsutum*) at its center of origin. *Molecular Ecology* 20: 4182–4194.
23. An excellent history of cotton cultivation and commercialization is presented in Beckert's *Empire of Cotton* (see note 10 for full reference).
24. The admission of Texas to the Union precipitated the Mexican-American War, which resulted in the further major loss of Mexican territories to the United States all the way to California.
25. By the middle of the 20th century, the cotton industry by and large moved out of Europe altogether.
26. The damming of rivers in this area and the use of their water to irrigate huge tracts planted with cotton led to the virtual disappearance of the Aral Sea in the early part of the 21st century.
27. Cigarettes were included until the 1970s in the food rations provided to US soldiers.

4 Killer Spices

THE SPICE TRADE UNTIL THE END OF THE 15TH CENTURY – THE RISE AND FALL OF VENICE, THE FIRST PREDATORY COMPANY MASQUERADING AS A STATE

Obtaining the basic food staples necessary for sustaining human life is the first step in the struggle for survival. Preserving food from spoilage and preventing the growth of human pathogens on foodstuffs are also important, and spices have traditionally been used to accomplish these goals. It is therefore not surprising that spices have been essential commodities for practically all human societies, and some spices – black pepper, clove, nutmeg, and, to a lesser degree, cinnamon (Figure 4.1) – have been at the center of major historical events, many of them violent.

Fresh plants or plant parts used in seasoning food and in cooking are considered herbs, and dry plants or plant parts used for similar purposes are called spices. For obvious technical reasons, until very recently only spices, but not herbs, were transported long distances, and this of course has allowed human societies to use plants that do not grow locally.[1] At least as far as Western Civilization is concerned, black pepper is probably the oldest spice of the four mentioned above, and the one consumed the most (by weight). The pepper plant, the species *Piper nigrum*, originated in the southwest corner of the Indian subcontinent, the area known as the Malabar Coast. Peppercorns were found stuffed in the nose of the mummified pharaoh Ramesses II, who died in 1213 BC, so we know that pepper was traded with the Middle East more than 3,000 years ago. It was well known to the Greeks, and after Nearchus, one of Alexander the Great's admirals, discovered the sailing route from India to the Persian Gulf, Greek ships regularly sailed the Red Sea to the Malabar Coast to trade for pepper. By the time the Roman Empire occupied Egypt, maritime trade between Rome and India via the Red Sea was substantial, with pepper constituting the bulk of the goods going to Rome, in exchange mostly for gold and other metals.

Pepper was consumed in ancient China as well. Chinese records indicate import of pepper from India as early as the 3rd century BC, initially probably by an overland route but later by ship. Eventually, pepper was exported to China also from the islands of present-day Indonesia, where pepper plants are believed to have been introduced by Indian migrants by the 1st century BC at the latest.

The spices clove, nutmeg, and cinnamon were also known to the ancient Greeks and to people both in the Middle East and in China. Clove is the dried flower bud of the tree *Syzygium aromaticum*, and the nutmeg spice is made from the dried fruit of the nutmeg tree, *Myristica fragrans*. The clove trees grew on just a few islands in the northern Maluku Islands region of present-day eastern Indonesia, and the nutmeg trees grew on a few islands in the central Maluku islands, about 500 kilometers south of the clove tree islands. Cinnamon is derived from the plant *Cinnamomum verum*,

FIGURE 4.1 Clockwise from top left: pepper, cinnamon, clove, and nutmeg plants. Pepper and nutmeg spices are prepared from the seed or seed cover, the cinnamon spice is obtained from the inner bark, and spice of clove is prepared from the dried flower buds.

which then grew on the island of Ceylon (Sri Lanka today), although some cinnamon traded with the West came from related *Cinnamomum* species growing in southern and eastern Asia, such as *Cinnamomum cassia*. However, trade with Europe in clove, nutmeg and cinnamon spices was always relatively smaller compared with the pepper trade, and clove, nutmeg, and some cinnamon had more distance to go and must have been transferred to India first on their way to Egypt or to other parts of the Middle East.

As long as the unified Roman Empire remained stable, trade with India for black pepper and the other spices proceeded smoothly. However, the split of the Roman Empire into Western and Eastern (Byzantine) Empires by the 4th century AD, the decline and fall of the Western Roman Empire, and finally the takeover by the Arab Caliphate[2] of the eastern shores of the Mediterranean Sea by the early 7th century AD from the Byzantine Empire had major consequences for the spice trade. Some amount of spices from the Far East, and other goods, notably silk, continued to arrive at Mediterranean Sea ports such as Beirut and Constantinople, the capital of the Byzantine Empire, via land caravans that used the reticulate route system called collectively "the Silk Road." The Arabs eventually picked up the maritime trade with India and the Far East for these spices. But the Arab states never developed strong navies that could protect cargo ships in the Mediterranean Sea from piracy, and controlling the lucrative last leg (lucrative for merchants and taxing authorities) in the trade of spices with Europe – the trip across this sea – was left to be contested by European states with maritime prowess. This struggle ultimately led to the rise of Venice, the first European empire chiefly engaged in, and vastly enriched by, the trade in the high-value commodities of spices, with pepper being the most commercially important among them.

Since the establishment of Venice as a republic in the 8th century, it could basically be characterized as an organization devoted to enriching its upright citizenry by commerce and looting, with preference for the latter. This statement should not be viewed anachronistically: looting and plundering on land, and piracy at sea, have been considered by virtually all peoples on earth as respectable activities at one time or another (present time not excluded). But certainly predatory Venice was a forerunner, and a role model, of many other capitalist-mercantilist ventures to follow, including several discussed in this book.

The conversion of Venice into a full-blown empire based on spices occurred gradually. Initially, individual Venetian traders sent ships to buy pepper, other spices, and silk from the Byzantine Empire, paying for them with slaves (mostly boys and young women) that they abducted or bought from Dalmatia, as well as salt, wood, grain and wine. The sale of the spices in European ports resulted in fabulous profits for these Venetian merchants. By the beginning of the 12th century, the Venetian Republic had perfected the process. The Republic's soldiers conquered and occupied multiple islands and coastal areas in the Adriatic and Aegean Seas and beyond (Figure 4.2). In its heyday, Venice occupied Crete as well as Cyprus, where Othello, admittedly a fictional character, was the governor on behalf of the Republic of Venice – to facilitate local looting and trading and to supply safe havens and provisions to the trading fleet on its way to Constantinople and other eastern Mediterranean ports. To support

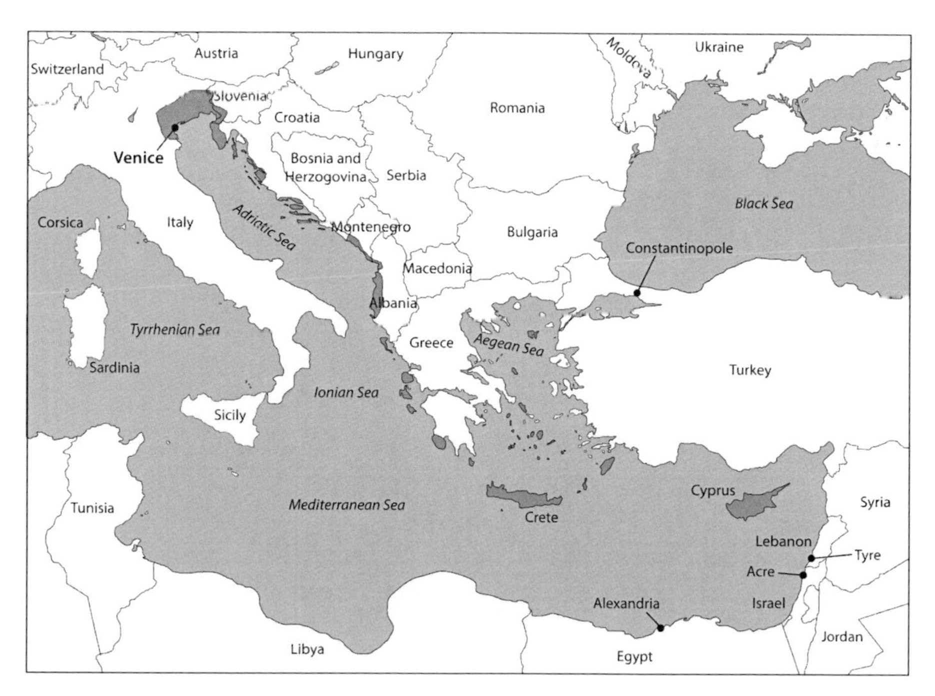

FIGURE 4.2 The territories occupied by the Republic of Venice at its peak, in the early 16th century (dark gray). Present state boundaries are shown.

these activities, the Arsenal in Venice was created (1104 AD) to build, at the government's expense, large ships that could carry both soldiers and cargo, and these ships could then be rented by one or multiple merchants, acting as a prototype "company", to engage in specific trading voyages.

In 1204, during the 4th Crusade and at the instigation of Venice, the crusader force sacked Constantinople, and much of the loot ended up in Venice. When the Byzantine Empire righted itself a bit and reasserted control over Constantinople and its vicinity, Venetian merchants were excluded from the city in favor of their Genoan competitors. In response, Venice established a military and trading post in Tyre (on the coast of present-day Lebanon). By the end of the 13th century, most of the ports in the eastern Mediterranean were controlled by the Muslim Mamelukes. Spices were now arriving from India and the Far East mostly by the sea route via the Indian and Red Seas to Egypt, and Alexandria became the main port for the trade of pepper and other spices with Europe. Venice dominated this trade by virtue of its military force, which it used to exclude competitors, chiefly the Republic of Genoa, from the sea routes.

The 15th century and the beginning of the 16th century saw the gradual expansion of the Turkish Ottoman Empire along the eastern coast of the Mediterranean.[3] The Turks took Constantinople in 1453 and they finally arrived in Egypt in 1517. The Turks repeatedly clashed militarily with Venetian forces, and once in control of all major eastern Mediterranean sea ports, they were not particularly favorable toward allowing Venice to trade in spices. Adding to Venice's problem, the expansion of the Turkish Empire coincided with the discovery of the sea route from Europe to the Malabar Coast of India around the African continent and the subsequent attempt by Portugal to corner the spice market (more to follow). While Venice still had access to the port of Alexandria and other ports on the east coast of the Mediterranean, its trade in pepper was cut by two-thirds at the end of the 15th century (from 1,600 tons per year to 500 tons/year), although by the middle of the 16th century it recovered somewhat. By then, however, the trade route for spices that went from the Far East to Europe via the Indian and Atlantic Oceans, skipping the Mediterranean Sea altogether, was well established, cutting into Venice's profits, and the Venetian Republic began its slow decline. Its demise came at the hands of Napoleon in 1797.

THE SPICE TRADE IN THE 16TH TO 18TH CENTURIES – THE RISE AND FALL OF VOC, A ROGUE STATE MASQUERADING AS A COMPANY

Much has been written about spices being the impetus for the Age of Discovery. The desire to find a shorter route to the Far East was motivated by the realization that the major component of the high cost of spices and other goods whose origin was the Far East was transportation, and therefore shortening the route and minimizing the number of times that the merchandize had to change vessels and ownership would maximize profits to the European traders. Portugal and Spain, two countries with ports on the Atlantic coast, were hoping to find such a route via the Atlantic and thus allowing their merchants to obtain spices cheaply and undersell the Venetians while still making a huge profit.

The Portuguese strategy was to attempt to reach India, believed to be a major source for spices (the exact source of the exotic spices was not known in Europe, although Marco Polo claimed to have visited some islands during his voyages in the Far East where such spices grew) by sailing the Atlantic around Africa. It took the Portuguese more than 50 years to figure out how to sail around the southern tip of the African continent and get into the Indian Ocean,[4] but once they did so, in 1488, it took them only ten more years to reach India, which they did with a small flotilla of ships commanded by Vasco de Gama.

The Spanish were also very interested in finding a short sea route to the source of these spices. Indeed, the reason the Spanish king sponsored Columbus' expedition in 1492 was exactly for that reason. By underestimating the circumference of the earth – and arguing against the consensus expert opinion which was in fact much closer to the mark – Columbus was able to convince the king that reaching the lands where spices grew would be faster by sailing westward. As is well known, Columbus died in 1506 without ever realizing that the land he had reached was not Asia. Columbus' calculation of the circumference of the earth was decisively shown to be a gross underestimation when a Spanish naval expedition (1519–1522) headed by the Portuguese navigator Ferdinand Magellan finally did reach the Far East by sailing west, going around the southern tip of South America. This expedition, which was the first to circumnavigate the planet, took three years to complete (Magellan was killed in the Philippines before the expedition reached any spice island), and it showed that the western route to the Far East was much longer than the eastern route.

At any rate, by that time the 1494 Treaty of Tordesillas, mediated by Pope Alexander VI, had divided the unexplored (by Europeans) world into West and East, giving the Spanish all future discoveries west of a meridian believed to run through the Atlantic Ocean, and the Portuguese all lands discovered east of this meridian. Curiously, while fully aware that the planet was a sphere, the Spanish and Portuguese negotiators of the treaty failed to discuss the border on the other side of the globe, allowing the Spanish to claim that the Far East "Spice Islands" were on the side of the globe that belonged to them. Indeed, this argument was the legal justification for the Magellan expedition in the first place. In practice, however, the Spanish efforts in the Far East were focused on their possession of the Philippines, using it mostly as a trading post with China, and they made only small incursions into spice island territory, as discussed below.

Once the Portuguese reached the western coast of India, they discovered that the main source of black pepper was its Malabar Coast,[5] and they began shipping pepper from there to Europe. They established military posts on the Malabar Coast and on the island of Ceylon, where the cinnamon tree grew, "claiming" parts of these areas for their king. These military campaigns, which were carried out in order to exclude or control the Arab, Malay, and non-Portuguese European merchants, eventually failed, because Portugal was never able to mobilize enough military power to accomplish this task. Nevertheless, the Portuguese exercise of military power in the Far East inflicted much cruelty and misery on the local population.

As part of this effort to have a monopoly on the spice trade, in 1511, soon after they established themselves in India, the Portuguese went around the Indian Subcontinent to occupy Malacca, which was then a Chinese protectorate, on the Malay Peninsula.

There they were told that the clove and nutmeg trees grew somewhere in the Maluku Islands archipelago, a region also sometimes referred to as the Spice Islands. The next year they sent an expedition that arrived at the Banda Islands, located at the center of the Maluku Archipelago, which were at that time the only place in the world where the nutmeg trees grew. In 1513 they arrived at a couple of other islands further north in the archipelago, Tidore and Ternate, which were covered with clove trees (Figure 4.3).

Arab traders had been to these islands before, and many of the locals had just converted to Islam. Tidore and Ternate were each governed by a Muslim Sultan, and the pair were competing with each other for local influence. The Ternate Sultanate was generally more powerful, controlling, albeit loosely, most of the territory and trade in the area, including the trade in nutmeg from the Banda Islands. Initially, the Portuguese began peaceful trading for clove and nutmeg with the locals, shipping the goods westward to Europe together with the pepper from India. However, the Portuguese, just like the Venetians before them, were obsessed with achieving a monopoly of the spice trade and so tried to force the locals to trade only with them and to exclude Arab and other European traders. Toward this goal, the Portuguese built forts on the islands and used their military to exert pressure on the local rulers. This of course was resented by the locals. In particular, while the Sultan of Ternate allowed the Portuguese to build a fort on his island in 1522, their constant interference in the political affairs of the sultanate caused repeated military clashes.

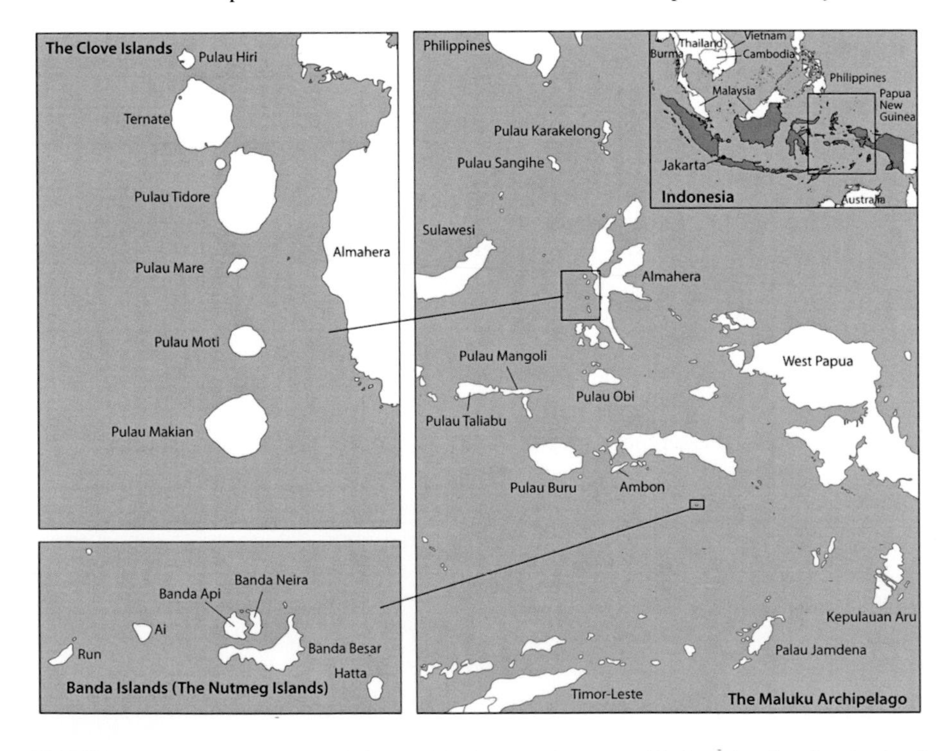

FIGURE 4.3 A map of the Maluku Archipelago, today part of Indonesia, showing in detail the original islands where clove and nutmeg trees grew.

When, in 1570, the Portuguese forces on the island executed Sultan Hairun and exhibited his head on a spike, the locals began a rebellion and finally expelled the Portuguese five years later. In the Banda Islands, the source of the nutmeg, the Portuguese also attempted to build a fort, in 1529, but were immediately thwarted from doing so by the locals.

The Portuguese were evidently not ruthless enough, leaving a wide opening for other European competitors. As mentioned above, the Spanish also made a few attempts to control these islands, starting with the Magellan expedition that visited Tidore. They made alliance with the Sultan of Tidore and later also invaded the southern part of Ternate. But they, like the Portuguese, could not resist for long the pressure from the Dutch, withdrawing for good from the area in 1663.

The Dutch incursion into the area, which proved long-lasting, began in 1581, when Holland became an independent country and in military conflict with Spain and Portugal. This conflict deprived Holland of its access to pepper, clove and nutmeg and other Far East spices, and it began sending its own merchant ships to this region. To prevent domestic competition and to have a stronger force in the Far East to face the Portuguese, the Dutch East India Company or VOC (from the Dutch *Vereenigde Oost-Indische Compagnie*) was formed in 1602, and given a monopoly on the spice trade by the Dutch government. VOC was the first permanent, completely limited liability company, and the first one to sell stocks. As a public company of course it operated for the sole purpose of making a profit for its stockholders, but its activities were not simply peaceful commerce. VOC acted as a state, with a government, navy, army, judicial courts, and the rights to issue its own coinage and to conclude treaties with foreign rulers. In essence, it was the new Venetian Republic, and for almost 200 years it plundered the Far East, to be eclipsed eventually by a similarly immoral organization, the British East India Company (EIC), which was established about the same time as VOC (Chapter 3, and see Chapter 5 for more of the EIC exploits in the Far East).

The VOC began by establishing permanent bases on the island of Java (in present-day Indonesia), with the main port at Batavia (now Jakarta), established in 1619, serving as a hub for its ships. The main commodities that VOC ships brought to Europe were pepper (which in addition to the Malabar Coast was also being grown in large amounts on Java and other Indonesian Islands), nutmeg, clove, cinnamon, and silk. Silk was obtained from China, using profits from the sale of pepper in China, where there was great demand for this spice. The VOC also engaged in trade with several other countries in the area, all the way to Japan, but always with the purpose of obtaining profits to buy spices and silk to sell in Europe.

From the beginning, Dutch methods antagonized both the locals and other European and non-European traders. The VOC ships were fully militarized, with naval guns and a detachment of armed soldiers. Since Holland was officially at war with Spain and Portugal, the VOC expedition commanders felt no qualms about engaging in warfare against naval and land forces from these two countries, and they also did not refrain from attacking the English, who were beginning to send trading expeditions to the area, as well as Malay and Chinese ships. As for dealing with the natives, the VOC standard approach was to get a local dignitary (or several dignitaries) to sign a long and complicated treaty, written in Dutch, giving the VOC sole rights in perpetuity to trade with them and often to occupy their land. Once such a treaty – which was not clearly understood, to understate the case, by the signers who, moreover, often lacked the authority to sign such a document – was signed, the Dutch were on the lookout for any perceived violation of the treaty, real or imagined. They typically had no problem finding such violations, and once they spotted them, they responded with full military force. As succinctly stated by the historian John E. Wills, Jr., the Dutch displayed "total indifference to legality and morality in [their] relations with Asian peoples,"[6] a statement that can justifiably be extended to include the conduct of all European powers then and for a very long time afterward.

Thus, from its establishment in 1602 to its formal demise in bankruptcy proceedings by 1800, the VOC engaged in military operations as much as in commerce, and in fact one of the major reasons for its ultimate failure was the cost incurred by the extensive military operations. For example, to provide a safe way station to and from Batavia, the VOC wrested Ceylon from the Portuguese in a series of military operations in the first half of the 17th century, and in 1641 they expelled the Portuguese from Malacca by force, thus ensuring safe passage for Dutch ships in the Straits of Malacca. Farther east, VOC controlled parts of the Indonesian Islands directly by force, although when possible, the company entered into military alliances with local rulers. And particularly at the beginning of VOC operation in the area, commandeering merchant ships of other Western countries as well as of local traders such as Malays, Chinese and Japanese and selling the captured goods was a common VOC practice, although it might be pointed out that all other European players in this arena also engaged in such activities.

Certainly VOC is most infamous for its behavior in the Banda Islands, the history of which has been often and extensively discussed elsewhere, so only the highlights will be mentioned here. Because this small group of tiny islands was the only place in the world where the nutmeg tree grew, the VOC was hell-bent on having complete control of the trade in this spice. By 1609, the Bandanese were so resentful of the Dutch's low prices and their attempt at military enforcement of an exclusive treaty with very bad terms to the locals that the Bandanese ambushed and killed about 40 Dutch officials. At the same time, forces of the British East India Company landed on two outlying islands, Ai and Run, built forts there, and were trading with the islanders, offering better prices and exchange goods than the Dutch cared to offer. Military skirmishes between the Dutch and the English in 1615–1616 on these two islands resulted in hundreds of casualties for both sides and the temporary retreat of the English. The Dutch soon dealt with the locals with even harsher measures. In 1621, the Dutch again forced local dignitaries to sign an unreasonable treaty, and after quickly perceiving that the treaty was not being honored, beheaded 40 of these dignitaries and displayed their heads on spikes. They then killed many of the locals (with the help of Japanese mercenaries) and expelled a large number of the remaining residents, with many others fleeing on their own.

Once the Dutch basically depopulated the islands of their natives, they began a plantation system run by Dutch settlers. These settlers, typically poor farmers in Holland who were promised the riches of the Far East, were supposed to run the nutmeg orchards with a workforce of about a thousand local residents who were forced to remain on the islands as slaves. But this system had a slow start since not many

Dutch farmers were interested in moving to the islands, and the VOC was forced to bring some of the expelled Bandanese, now working as slaves in Batavia, back to the island, as their knowledge in tending the nutmeg trees was essential. More slaves from elsewhere in the Far East were also brought to the island to work on the plantations.

The brutal treatment of the local residents and the complete military occupation of the islands by the Dutch, and the tree eradication program that the Dutch undertook in outlying islands where they could not easily exert control, ensured that no nutmeg would be sold to anyone without Dutch being the intermediaries. However, following the 1615–1616 military skirmishes on Ai and Run between the VOC and the EIC forces mentioned above, in 1619 the rulers of the two countries forced the VOC to agree to peaceful co-existence with EIC, in which VOC controlled two-thirds of the trade and the EIC one-third, and the English were allowed to reoccupy their fort on Run. The local headquarters of the two companies were placed in Ambon, an island just north of the Banda Islands that was governed by the VOC governor and whose port served as a collection center for various types of spices obtained from the region, including nutmeg, clove and pepper.

But in 1623, the Dutch authorities claimed that they uncovered a conspiracy by the English to seize the local Dutch fort and kill the governor. They obtained confessions from several Englishmen by employing the waterboard torture technique, after which they tried them for treason, found them guilty, and executed ten Englishmen, nine Japanese (Japanese ronins – freelancing samurais – worked as mercenaries for the VOC) and one Portuguese by beheading, again displaying some of the severed heads on spikes. The English came to call these executions the Ambon Massacre.

This incident caused a new rift with the English, and led directly or indirectly to three successive Dutch–English wars that were fought mostly elsewhere in the world, only the second of which had a major impact on the spice trade. That war was concluded by the Treaty of Breda in 1667 in which the island of Run was officially exchanged for the island of New Amsterdam (now called Manhattan) in North America, basically confirming the facts on the ground, since the Dutch were already in possession of Run at the time and the British held Manhattan. So while the EIC had mostly withdrawn from the Maluku Islands already following the Ambon Massacre, preferring to concentrate on its trade with India, the Breda Treaty ensured the complete control of VOC over the nutmeg trade until the company's demise in 1800, which coincided with the Napoleonic (or Revolutionary) Wars. Around the time of the Breda Treaty, the Dutch also consolidated their control over the entire Maluku Islands, by dint of military force and treaties with the local sultanates, thus completing the exclusion of the Portuguese and Spanish from the area and ensuring their monopoly on the clove trade as well.

These monopolies came to an end during the Revolutionary Wars. Until then, while attempts to smuggle nutmeg and clove plants out of the Maluku Islands were constant, the Dutch were diligent in combatting those efforts by both sterilizing exported nutmeg (which is, botanically speaking, the seed) as well as raiding all unauthorized orchards and killing the plants and, often, the farmers. Some attempts at smuggling plants, notably by the French in 1770, were successful, but the trees did not do well in their new home, Madagascar. However, during most of the

Revolutionary Wars, starting a few years after the French Revolution in 1789 and lasting until about 1812, Holland was occupied by French forces, and since France was at war with England, the English felt legally justified in invading and occupying the Banda Islands. They did so twice during that period, and while eventually they returned the islands to Dutch control, they withdrew from the area only after they had removed both nutmeg and clove plants and replanted them in Ceylon, which was by then under English control, and in other English colonies in the tropics. Eventually, clove and nutmeg trees came to be cultivated in much larger areas around the world than in the Maluku Islands, including places with better and closer access to Europe, making possession of these islands or even trade with them not particularly profitable.

THE ATTRACTION OF SPICES TO HUMANS – IS IT CHEMISTRY OR METACHEMISTRY?

There is general agreement among historians that the proximate cause of the aggressive, militaristic, and immoral behavior described above was the fact that engaging in such activities was highly profitable, or at least perceived to be profitable. A biologist would add, as stated in previous chapters, that money is a proxy for fitness – the more money one has, the more fecund one is likely to be and better at ensuring the continuing success and multiplication of one's progeny, which is the definition of biological fitness. In economic terms, it is also clearly understood that this profitability was due to high demand and limited supply. The disagreements begin when trying to explain the demand. Many theories have been developed,[7] and some will be summarized below. However, much of this debate was concentrated on human cultural aspects, and European culture specifically, and argued by scholars whose expertise is in this area rather than in biology. While some of these arguments must have some validity, it is my contention that once we consider information about the plants themselves, their chemical compositions, and the biological effects that these chemical have on humans and other organisms, we are likely to develop a much better understanding regarding the ultimate reasons for human use of, and desire for, spices throughout history. Luckily, recent studies have provided us with a wealth of such information.

Most people would agree with the statement that spices make food tastier, at least when done right. But the argument that we like to eat spiced food because it tastes good is a proximate explanation, not the ultimate one. Evidently since natural selection has resulted in people who find the consumption of spiced food to be a positive, pleasant sensation, there must have been (and maybe there still is) an adaptive value to liking and enjoying spices. But what is this advantage? This question is even more pertinent in view of some of the toxic and noxious properties of many of the chemicals found in spices and which are outlined below. I will describe some likely possible advantages later on, but first let us discuss some of the less likely explanations why people consume spices.

One possible notion was that, at least in the past, spices were simply used to help disguise the bad taste and smell that emanate from spoiled food, particularly spoiled meat. This notion is now generally dismissed, and rightly so. Even if spices were capable of overcoming the noxious aromas of rotten meat – and if this is all they could do – anyone who found such food appetizing to consume would not likely survive the experience, most likely succumbing to food poisoning.

Having dismissed this notion, most commentators continue by arguing that the proposition that spices, many of which have recently been shown to have antimicrobial properties in controlled laboratory tests, were used to preserve food is also not correct because the amounts of spices applied to the meat was insufficient to achieve such a goal. They further argue that good quality fresh meat was not difficult to come by in the past, particularly for the rich, who were the main consumers of spices (and meat). Alternatively, the argument is made that although fresh meat might have been unavailable during the winter (since culling of the herds was done in the fall, so they would not have to be fed with stored hay during the winter), drying, salting, and smoking were cheaper and more efficient ways of preserving food. Once such simple "biological" explanations are dismissed, the commentators concentrate on the often ostentatious consumption of spices by the nobility and rich burghers throughout European Medieval history, and propose a variety of explanations, from simply the flaunting of wealth to the spiritual or symbolic meanings of spices perceived by these consumers. They then go on to describe a "collapse" in the demand for spices in Europe toward the end of the 18th century, due to various changes in social thinking and behaviors, which put an end to the "Spice Age".

While the spice habits of rich Europeans might tell us something about their culture and values, the emphasis on studying the demand for spicing by a small segment of society in a single geographic region of the world misses, in my opinion, the main explanation for the use of spices by humans in general. To begin with, spices have been used continuously in practically all human societies around the world since antiquity, and by all segments of these populations, including poor people in Europe during the Middle Ages. While the use of specific spices depends of course on their availability, we will see that people prefer spices that contain specific chemicals with important biological effects. Second, there are no data that show that, once the small numbers of Europeans who clearly engaged in conspicuous consumption of spices are taken out of the equation, the per capita consumption of spices – in Europe or elsewhere – has declined over the years. Indeed, it appears that pepper and cinnamon are ingredients found in many types of foods today, including ready-to-eat foods, and while present use of clove and nutmeg is not as extensive as black pepper is, they never were used heavily by the general public. Moreover, substitute spices that contain similar compounds are globally prevalent in modern cooking, like allspice (*Pimenta dioica*, a New World species that is closely related to the clove tree) as well as basil (*Ocimum basilicum*), which is technically not a spice but a herb, both of which contain eugenol, the main chemical found in clove. So the drop in the prices of spices was probably due simply to increased supply of most spices, and the adoption of chemically similar substitutes rather than to decreased demand. And with the increase in world population, overall production of most spices continues to increase. For example, worldwide production of black pepper reached 350,000 metric tons in 2012, the highest ever.

The constant and universal demand by humans throughout history for spices in their food suggests that there must be biological reasons for this demand.

A commonly made observation is that countries located in warmer climates tend to have more spicy food. To get quantitative data regarding spice usage, Prof. Paul Sherman and colleagues analyzed extensive data on use of spices in different countries and showed that there is in fact a strong correlation with the average temperature in a given country and the number of meat dishes that are spiced, as well as the number of spices used per dish.[8] And contrary to some previous reports, they also showed that many spices are quite efficient at killing many microorganisms that spoil food, as well food-borne bacterial pathogens, at spice concentrations used in cooking the dish. Furthermore, other research has shown that most of the spices keep their microbicidal properties even after the cooking process. Moreover, Prof. Sherman's research showed that vegetarian dishes, which spoil more slowly than meat, tend to be spiced less during preparation.[9] Thus, it appears that the chemicals in spices can indeed prevent food spoilage, and thereby prevent people from getting sick. But why do plants make these compounds in the first place?

The part of the black pepper vine plant that is used as the spice is the seed, often referred to as the peppercorn or the berry. The main compound found in the seed of black pepper is piperine, and two other major constituents are limonene and sabinene.[10] Depending on how the seeds of *Piper nigrum* are collected and treated, the spice might be called black, white or green pepper; however, there are practically no differences in the composition of the main active chemicals in these different types. Therefore, the term "black pepper," the species name, is typically used to refer to all of them, and to distinguish this spice both from the closely related species, *Piper longum* ('long pepper'), which was used extensively in antiquity and also contains piperine, and chili pepper, which is a general name for several species in the South American genus *Capsicum* that are not closely related to black pepper and that do not contain piperine.

In clove spice, prepared from the dried buds of the tree *Syzygium aromaticum*, the main compound is eugenol, with β-caryophyllene also present.[11] [In an aside, generic chemical names are typically given by the person who first isolates and identifies them, and are often based on the original source of the chemical. The clove tree was once called *Eugenia aromaticum*, so the main compound found in clove spice was called eugenol. Sometimes, as is the case with eugenol, it is later discovered that many other plants are capable of making the same compound]. The same main active chemicals are found in clove's close relative allspice,[12] although in this case the spice is prepared from dried fruits. As for the nutmeg tree, there are actually two spices obtained from it. The nutmeg fruit has a seed inside it, and the seed coat ("shell") is covered by a thin, red, perforated membrane called *mace* (Figure 4.4). Mace spice is prepared from the dried and ground mace, while the dried and ground seed is used for the spice called simply nutmeg. Both nutmeg and mace spices have similar compositions of the active chemicals, myristicin, elemicin, and sabinene being the three most abundant.[13] However, these chemicals are found in a purer, more concentrated form in mace, since the seed itself has many other compounds that mix together with the active ingredients when the seed is dried and ground to make a powder. The main chemical in the cinnamon spice, which is the dried and ground inner bark of young shoots, is a compound called cinnamyl aldehyde, with eugenol a distant second in abundance.[14]

FIGURE 4.4 A clove tree with buds (top left), picked clove buds drying (bottom left), nutmeg tree with fruit (top right), and a nutmeg fruit from which the outer layer has been partially removed, showing the red mace and the dark brown nut inside (bottom right). Pictures were taken in 2015 on the islands of Ternate and Tidore, Indonesia. Clove is indigenous to these islands, but nutmeg is a recent introduction.

The main active compounds found in black pepper, clove, nutmeg and cinnamon fall into two chemical classes (Table 4.1, Figure 4.5). The first group of compounds, called "aromatics" because they all have a phenyl ring[15] as part of the compound, includes eugenol, myristicin, elemicin, cinnamyl aldehyde, and piperine (although piperine is also an alkaloid, meaning that it contains a nitrogen atom; see Chapter 5 for more extensive discussion of alkaloids). Limonene, sabinene and β-caryophyllene belong to the second class, called "terpenes," which are defined as compounds having carbons in multiples of five (the five-carbon unit is called an "isoprene") and a ratio of eight hydrogen atoms to five carbon atoms (i.e., $(C_5H_8)_n$, n being any integer from one onward), and with an oxygen atom sometimes present as well.

Recent studies have shown that all these compounds, just like nicotine discussed in the previous chapter, have toxic effects on microorganisms and animals. A standard test for antimicrobial activity involves growing various bacterial or fungal species on a series of petri dishes that contain, in addition to nutrients required for the growth of the microorganism, the tested compounds at different concentrations. After spreading a fixed number of bacteria or fungal spores on the plate and allowing them to grow for a few days, the number of bacterial or fungal cells that managed to grow on each of the plates are counted. In such studies, all of the compounds mentioned above have proven toxic, at relatively low concentrations, to many microorganisms.

TABLE 4.1
The Primary Active Chemical Compounds in Black Pepper, Clove, Allspice, Nutmeg, and Cinnamon

	Aromatics	Terpenes	Is the Spice "Hot"?	Plant Part Used
Black pepper	Piperine	Limonene, sabinene	Yes (piperine)	Immature fruit
Clove	Eugenol	β-caryophyllene	No	Bud
Allspice	Eugenol	β-caryophyllene,	No	Immature fruit
Nutmeg (including mace)	Myristicin, Elemicin	Sabinene	No	Seed, seed coat
Cinnamon	Cinnmyl aldehyde, Eugenol		No	Inner bark, young shoot

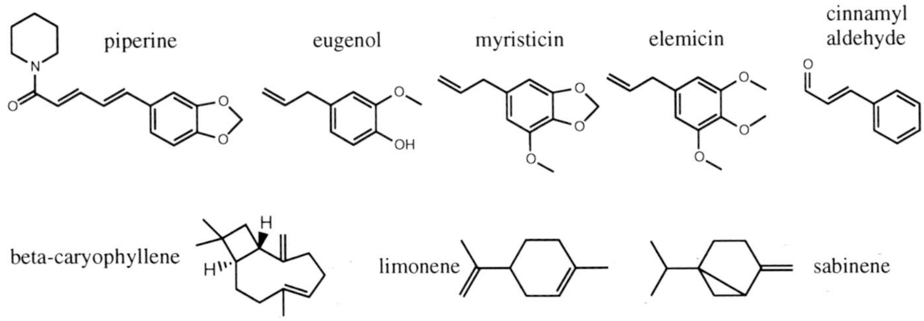

FIGURE 4.5 The structures of the compounds listed in Table 4.1. Top row from left: the aromatic compounds piperine, eugenol, myristicin, elemicin, and cinnamyl aldehyde. Bottom row from left: the terpene compounds β-caryophyllene, limonene, and sabinene.

These results are consistent with traditional observations, for example that the addition of clove to dough prevents or greatly delays mold from growing on the baked goods. These compounds are only mildly toxic to animals, as shown by assays in which typically mice or rats are fed these compounds at different concentrations. Eugenol, myristicin, cinnamyl aldehyde and β-caryophyllene have LD_{50} values (see Chapter 3 for the definition of the term LD_{50}) that range from 1,900–5,000 milligrams (mg) for mice or rats, although myristicin has an LD_{50} value of 400 mg for cats. Piperine, which is the least toxic to bacteria among these chemicals, has the highest toxicity of this group of chemicals to mice, with an LD_{50} of 400 mg.

Each animal species might exhibit a different sensitivity to a given compound, as is demonstrated by myristicin,[16] and these different sensitivities are also true for microorganisms. The high sensitivity of a species to a compound might indicate that there is a specific "target" in cells of that species that the toxic compound hits, or that the species is particularly weak at deflecting or destroying (i.e., detoxifying) this compound. However, the observation that all of these compounds are toxic to some extent to many disparate living organisms may mean that something common in their structures is responsible for this toxicity. In fact, there are two general properties that are shared by all these compounds of both the aromatic and terpene groups. One is that they are rather oily substances, meaning they do not mix with water very well (they are said to be hydrophobic), but they mix well with oil and other organic liquids, including alcohol. Indeed, spiced wine is an old tradition. The second is that they have an abundance of carbon-carbon double bonds, meaning that they have two adjacent carbon molecules that are linked to each other by two covalent bonds, indicated as C=C.

When compounds that are not water-miscible come in contact with a living cell, they typically end up in the oily cell membrane,[17] where many proteins important for the proper function of the cell are located. If the concentration of these hydrophobic spice chemicals is sufficiently high, they might interfere with the physical properties of the membrane, which is made mostly of fatty acids, and therefore indirectly interfere with the proper working of the membrane proteins. However, the stronger toxic effect of these spice chemicals is probably due to the presence of the carbon-carbon double bonds. One of these bonds can easily shift to another atom nearby. This often results in chemical linkage of these compounds to a variety of important cell molecules, such as proteins or metabolites, rendering these cellular molecules abnormal and non-functional. In addition, when an oxygen atom is nearby, the bond-shifting of the spice molecule could link it to oxygen, an outcome in which the spice molecule is said to be "oxidized." Spice molecules with carbon-carbon double bonds are typically strong "antioxidants," because by being oxidized themselves they remove oxygen from the vicinity and thus prevent other compounds from being oxidized. Since oxygen is required for growth of many living organisms, removing oxygen from the environment often inhibits or slows down the growth of many microorganisms. The antioxidant activity of spices is probably the original reason for adding them to wine, to prevent spoilage by microorganismal growth.

However, while cell lethality or at least cessation of growth in organisms that harm plants are in a sense the "best" outcomes for the plant in its effort to combat both its microbial and animal enemies, there are many other negative effects that these compounds can cause, just as we have seen with nicotine in the previous chapter. Eugenol is known to be a neurotoxin, and in fact dentists use it to this day as an analgesic by rubbing it on sore gums. Ingestion of myristicin causes hallucinations in humans at a mere concentration of 6–7 mg per kg body weight.

While the molecular mechanism of myristicin intoxication has not yet been elucidated, the reason why piperine in black pepper causes the "hot" sensation is now well understood. Piperine binds to a protein receptor called TRPV1 on the surface of a specific type of nerve cells.[18] Once bound, the neuron fires, and the person experiences a painful, burning sensation. The same neuron also fires, without piperine present, when the temperature rises to 43°C. Thus, this is really a neuron whose function is to detect excessive heat. Interestingly, TRPV1 also binds capsaicin, a molecule found in chili pepper, even more efficiently than it does piperine, so consuming capsaicin gives a more painful sensation than piperine causes when applied at equal concentrations. A look at the two-dimensional structure of piperine and capsaicin reveals that they are both *amides*, meaning that each has a carbon that is bound to both a nitrogen atom and an oxygen atom (Figure 4.6). They also both possess a benzene ring to which additional atoms are attached.

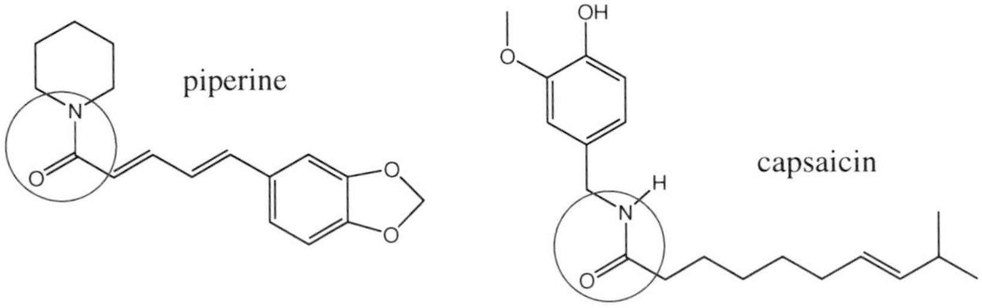

FIGURE 4.6 Comparison of the chemical structures of piperine and capsaicin, the active chemicals in the "hot" spices black pepper and chili pepper, respectively. The red circles show the amide group in each compound.

The modified benzene ring and the amide group have been shown to be involved in the binding of capsaicin to TRPV1, and the same is likely to be the case for piperine as well.[19] The binding of these two sections of either spice molecule to the much larger proteinaceous TRPV1 receptor molecule occurs because the contours (i.e., three-dimensional structure) of these sections are complementary to the contours of a particular region in the TRPV1 receptor, thus allowing a spice molecule to come in close contact over an extended surface area with the surface area of this receptor molecule. Once in close contact, several molecular attraction forces help keep the spice molecule and the receptor "bound" to each other. The mechanism of binding of a small molecule (often referred to as a "ligand") to a specific region of a large protein receptor by means of complementary surface topologies has been likened to the way a key fits into a lock.

An interesting observation is that birds, which are the seed dispersing agents of both black pepper and chili pepper, lack the TRPV1 receptor and can therefore eat the fruits (containing the seeds) of these two plants with impunity. It therefore appears that the ability to make piperine and capsaicin in these plants evolved as an adaptation to prevent their fruits from being eaten by the wrong animals.

Black pepper, with its pain-causing piperine, has been used as a spice since antiquity in the Old World, and so has capsaicin-containing chili pepper in the New World. In fact today black pepper and chili pepper are overall the two most often used spices anywhere in the world, as judged by the percentage of recipes in which they are included.[20] So, why has black pepper been such a popular and ubiquitous spice, especially since its antimicrobial activity is not particularly strong? One reason is that piperine actually does have specific and highly efficient toxicity to a ubiquitous and deadly food-borne bacterium, *Clostridium botulinum*. Second, the presence of piperine enhances the gut absorption of other spice chemicals, which fits with the observation that black pepper is often used with other spices. For example, French sausages are often made with a spice mixture called "quatre epices" that contains black pepper, clove, ginger and nutmeg. This is noteworthy since sausages are highly susceptible to *Clostridium botulinum* bacterial cells, which thrive and multiply in such rich source of food.[21] Growing *Clostridium botulinum* bacterial cells secret botulinum, a deadly neurotoxin to which humans are extremely sensitive (LD_{50} = 2 ng/kg).

Perhaps even more importantly, the durability of black pepper as a spice may have to do with its affinity to the TRPV1 receptor. Besides being a heat sensor, this receptor is a major cause for sensation of pain in inflamed areas. Piperine happens to be very good at binding and thereby desensitizing the TRPV1 receptor, while eliciting a lot less pain than capsaicin does. Given that in the past many people suffered from poor health of their gums and teeth, and some people still do today, the lasting popularity of pepper with its pain-relieving piperine and clove with its gum-numbing eugenol is perhaps not surprising. The effect of the mild hallucinogenic myristicin in nutmeg could only add to the overall positive contributions of these spices to human nutrition and to feelings of well-being generally associated with eating, on top of their antiseptic properties. With such psychopharmacological effects, it is not surprising that cases of human addictions to many spices have often been reported.

It therefore appears that the toxic chemicals that plants make for their own defense are used by humans in the same way – to kill microorganisms and small animals such as worms and other insects that have found their way into our food. While such compounds are toxic to us humans too, we put just enough of the spices in our food to harm our own enemies but not ourselves. In addition, while some of these compounds are neurotoxic, we use just enough of them to self-medicate ourselves to achieve a desirable psychopharmacological effect for us while generally avoiding the bad psychopharmacological consequences that would come from heavier consumption of spices.

While the beneficial effects of the consumption of spices can be documented, the origin of the practice of using spices in our food is still not obvious. Given that spices contain toxic chemicals, and they often impart strong flavors that one has to get used to, how did humans in their past history evolve the behavior of using spices in their food in the first place? Why do most of us today prefer, when given the choice, to eat spiced food? Is this behavior determined by our genes, was it completely culturally acquired in the first place and is now culturally transmitted, or does the explanation involve a bit of both?

To explore these questions, for which there are still no definitive answers, we need to start by discussing how we detected the active ingredients in spices. It turns out that with few exceptions, notably piperine, we actually detect them with our nose, and not in our mouth, as people often assume. As described in Chapter 3, the human mouth is equipped with just five specific detectors for the sweet, salty, bitter, sour and savory (umami) tastes. Detection occurs using in principle the same lock and key mechanism described above. For example, as described in Chapter 3, the taste buds responsible for detecting sweetness have nerve cells whose surfaces are lined with protein receptors whose own surface contours are complementary to the shape of certain chemicals, such as sucrose. Consequently such compounds, when present in the food, will bind to the sweet receptor and when binding occurs the neuron will fire and the brain will interpret this firing as an indication of the presence of a sweet chemical. Likewise, there are receptors for bitter compounds, and salty, sour and savory compounds as well. However, with only five receptors with very distinct binding capabilities, the range of sensations is not large. Basically, any compound that binds to the sweet receptor is simply experienced as a sweet chemical, and there are hundreds of compounds that bind to the same sweet receptor, because, although they are overall different from each other, they still all have a part that is similar to each other and is complementary in shape to the receptor. Thus they all cause the neuron to fire and as a consequence we cannot tell the compounds apart (the after-taste of some artificial sweeteners is due to the fact that they also bind, albeit not very strongly, to other taste receptors in the mouth).

But we can easily tell clove, cinnamon, and nutmeg apart, and for that matter whether we are eating a peach or a plum, a steak or a lamb chop. We do that through our nose, which is capable of detecting the multitudes of flavors that our foods possess besides the five basic tastes enumerated above.[22] The olfactory system in our nose has neurons with receptors too. But instead of just five receptors found in the mouth, there are perhaps as many as 400 different types of receptors, and each neuron cell in our nose contains only one type of receptor. Again, each receptor has a certain shape, and it can bind a compound that has a complementary shape. Once bound, the neuron will "fire" and a signal will arrive at the brain. While it appears that humans should be able to tell only 400 different odors, in fact we are capable of detecting a lot more. This is so because many odor compounds can bind to more than one scent receptor, usually with different parts of the molecule (if the compound is not capable of binding to any receptor, then we simply cannot smell it). So, for example, odor compound X might bind to receptors of type "23" found in one neuron, and to receptors of type "324" found in another receptor, and odor compound Y might bind to receptors "17," "121," "128," and "324," each present on a different neuron, and so on. Thus, the number of permutations is enormous, and the firing of each set of neurons constitutes a specific pattern, or "signature," that we experience as a different sensation.

There are air passages leading to our olfactory organ in the nose directly from the outside (our nostrils), as well as from the back of the mouth, the so-called retronasal passage. At any rate, to reach the odor receptors the odor compounds need to be airborne. This is simple enough for material that is gaseous at ambient temperature. But what about all the spice compounds as well as myriad other compounds in our food that give our foods its distinct flavor? Eugenol, for example, is a compound that is an oily liquid at temperatures below 254°C, and cinnamyl aldehyde, myristicin, β-caryophyllene, sabinene, and many other flavor compound found in spices and regular foodstuffs are similarly liquid or even solid at ambient temperature. What makes them flavor compounds is their ability to easily volatilize, or evaporate, at temperatures even below their boiling point, so that at least a few of these molecules become airborne at such temperatures.

There are a couple of physical characteristics that lead compounds to evaporate easily. First, they need to be small enough. Second, compounds that are hydrophobic evaporate better. This is because foodstuffs always contain some water, and compounds that are miscible in water are then mostly dispersed throughout the water volume, where they are also weakly bound to the water molecules. Such molecules, in order to become airborne, need first to reach the surface of the water, and this has to be accomplished by randomly traveling (i.e., diffusing) throughout the body of water while being constantly slowed down by the weak attraction forces of the water molecules. Molecules that do not mix with water tend to be pushed outward toward the surface of the water and can thus easily escape into the air.

And in general, the warmer the temperature, the faster the rate of evaporation is for any chemical.

Eugenol, cinnamyl aldehyde, myristicin, elemicin, β-caryophyllene, limonene, and sabinene are all hydrophobic compounds that are small enough to evaporate easily. They thus fall into a group of compounds we call volatile organic compounds, or VOC (yes, another VOC). This is why, when we put our nose to a spice vial containing clove or cinnamon or other spices we can smell them immediately. We can even smell food dishes, or ripe fruits, simply by putting our nose right next to them, or, if they are particularly redolent, we can smell them from some distance. But even more consequentially, when we put food in our mouth and chew it, some of the molecules of such compounds present in the food that are on the surface of the food particle will evaporate and then reach the olfactory organ through the retronasal passage. When the food is warm or even hot, such evaporation will be even faster, and at any rate masticated food in our mouth quickly warms up by the heat of our mouth. Our olfactory neurons will therefore instantly bind the VOCs, and we will simultaneously experience the sensations perceived by our taste buds in our mouth and the smell of the food volatiles in our nose and integrate these separate inputs into a completely seamless sensation (the "feel" of the food in our mouth also exerts an influence of what we come to regard as the unique flavor of the food). Since each natural food or prepared dish inevitably has multiple VOCs present in unique proportions, we perceived each food as having a unique aroma. Most of us are apparently not aware that our experience of flavor requires our nose – we think it comes from our mouth because that's where the food is – but the role of our nose becomes readily apparent when we have a cold and our nose is stuffed and suddenly our food seems to have no flavor.

This is how we recognize the presence of spices in our foods, and smell plays an extremely important role in this process. However, not all chemicals the general public consider as spices are volatile. Notable exceptions are piperine and capsaicin, whose molecules are too large to be volatile, but, as we have seen, they are recognized by the TRPV1 heat receptor.

But let's return to the question of why we generally like spiced food and find, sooner or later, the flavor imparted by the VOCs in spices attractive. The sense of smell is surely the first sense that evolved in living organisms. Even bacteria can sense chemicals in their environment via specific binding of such chemicals to receptors in or on the surface of the cells, and incidentally plants can sense VOCs as well. The ability to sense a chemical in one's environment – be it air or water – is crucial for a successful life. It is not surprising that we are capable of smelling smoke from fire and recognizing it as a danger sign. We are also very sensitive to VOCs that emanate from spoiled meat (such volatiles tend to contain nitrogen) and from spoiled vegetables (these volatiles contain sulfur), because avoiding consumption of such spoiled food is clearly beneficial to our survival. So while our ability to detect such compounds has been meticulously documented and measured – for example, we can smell ethyl mercaptan, which comes off rotten plants, at a concentration of 0.00000066 mg per liter of air – there is still some disagreement among the experts on whether our avoidance of such compounds is innate or learned. One opinion is that we are genetically programmed to dislike such compounds and so avoid food containing them, while another opinion is that our dislike of them develops only after we first eat spoiled food, get sick, and learn to associate the bad experience with these VOCs. It is even possible that we learn to dislike these VOCs culturally, when, for example, our parents point out the smell and tell us to avoid it.

As with our response to the smell of rotten food, similar arguments have been brought forth regarding human preferences for spices. Do we have innate, and unconscious, preference for spiced food, or is this preference strictly learned and culturally transmitted? Clearly, we have the machinery to smell the VOCs in spice, and therefore there must have been selection to evolve and maintain the necessary receptors and the neural circuitry, just like there has been selection to evolve and maintain the receptors to smell rotten food. In fact, recent results indicate that, contrary to prevailing notions, humans possess a generally very keen sense of smell, and particularly high sensitivity to many odors, some of which we can detect at much lower concentrations than many other animals are incapable of detecting.[23]

It seems a bit unlikely that selection was strong enough to favor the evolution of such numerous and effective scent receptors but not the evolution of an innate response of dislike (in the case of rotten food) or like (in the case of spices). Certainly in other cases, such as the sensing of sweet chemicals in humans, evolution has done both, and babies are born with an ability to taste sweetness and a preference for it. What complicates the argument for innate preference is that this preference does not appear immediately upon birth. Young children naturally avoid spicy food and plant foods that are rich in these toxic chemicals that are the essence of spices. But so do teenagers going through puberty and women in the first trimester of pregnancy. What these three groups have in common is that they all represent individuals who have some cells that are growing and dividing fast and are therefore most susceptible to the toxic effects of these chemicals. What this observation does suggest is that the avoidance of spiced food by these groups is unconscious. And if avoidance is an evolved innate response, there is no reason why preference for spices could not also be an evolved innate response, although one that comes into effect a few years after birth and that is programmed to temporarily disappear during puberty and early stages of pregnancy, only to reappear when the vulnerability to the chemicals in the spices abates.

Under this scenario – again, still only a hypothesis – individuals who either were genetically unable to detect spiced food because they lacked the receptors, or who could detect the VOCs but had no genetic disposition to prefer such food, tended to eat non-spiced food more often than those people who could detect spices, and were therefore more prone to suffer food poisoning and early death. Thus, natural selection favored those people who could detect spices in their food and favor it over non-spiced food. It should be pointed out that such selection could have started before people learned how to cook and otherwise prepare elaborate dishes, and probably involved natural plant foods, and combinations thereof, that either had or did not have such toxins and therefore had different lengths of "shelf-life" before they began to spoil. Only later did people start to add these plant materials to the food that they stored and cooked, possibly by finding the advantage of doing so by chance, and eventually to culturally transmit this custom. Given that there are differences in spice use between different human societies living in tropical, subtropical and temperate climates even today,[8] comparing the scent receptor genes of humans in these societies for evidence of stronger positive selection in people living at warmer regions might lend support to the hypothesis that human spice use has been favored by natural selection.

CONCLUSIONS

Existing evidence suggests that we humans have evolved to like spiced food, and that the consumption of spiced food has a beneficial outcome on human health. Regardless of whether people have been conscious of this benefit or not, they have sought to acquire spices to use in preparing their food. Consequently, when certain spices were in short supply, particularly in the more temperate zones of the world such as Europe, their prices were correspondingly high. In turn, the opportunity to make enormous profits in the spice trade has caused many people, separately and in groups, to engage in various actions that have ranged from peaceful trading all the way to violent acts and large-scale wars that have brought enormous misery to huge numbers of people around the world.

NOTES

1. For general historical background on human use of spices and the spice trade, please see Czarra, F. 2009. *Spices: A Global History*. Reaktion Books Ltd.; Milton, G. 1999. *Nathaniel's Nutmeg*. Penguin Books; Freedman, P. 2008. *Out of the East: Spices and the Medieval Imagination*. Yale University Press; Schivelbusch, W. 1993. *Taste of Paradise: A Social History of Spices, Stimulants, and Intoxicants*. Vintage Books; Turner, J. 2004. *Spice: The History of a Temptation*. Vintage Books.
2. Lapidus, I. M. 1988. *A History of Islamic Societies*. Cambridge University Press.
3. Lord Kinross. 1977. *The Ottoman Centuries: The Rise and Fall of the Turkish Empire*. Morrow Quill Paperbacks.
4. A riveting description of the Portuguese expeditions is found in Crowley, R. 2015. *Conquerors: How Portugal Forged the First Global Empire*. Random House. The Portuguese navigators faced two major hurdles. The first was that just south of the equator the winds regularly blow in a westerly direction. The second was that on long sea expeditions – a hallmark of the Age of Discovery – sailors invariably began to suffer from the debilitating disease of scurvy, which eventually killed them. The navigation problem was solved by the 1488 Bartholomeu Dias expedition with the creative solution of tacking southwest until the 38° parallel south, then catching the Westerly winds and sailing due east (this maneuver led to the chance discovery of Brazil by another Portuguese expedition in 1500, see Chapter 3). The solution to scurvy took much longer to resolve. Scurvy is caused by the lack of the chemical ascorbic acid, also known as vitamin C. This chemical is required for the activity of several enzymes that catalyze essential metabolic steps, both in animals and plants. While many animal species (and all plants) are capable of synthesizing ascorbic acid from common organic molecules, *Homo sapiens* is not one of them, and humans need a constant supply of it in their diet. Some plant foods, for example citrus fruits, are particularly rich in vitamin C. Sea captains were often aware that sailors suffering from scurvy quickly recovered upon eating fresh fruit available at ports (although ascorbic acid itself was not identified conclusively until the early 20th century). While the Portuguese relied on local inhabitants along the coast of Africa to supply them with fresh fruits, the Dutch East India

Company, once it entered the Far East trade, established a colony of Dutch farmers in 1652 at the present location of Cape Town in South Africa to supply their ships with produce. This colony was the first European settlement in South Africa.

5. They were not the first Europeans to discover it. Marco Polo and Nicolo de Conti, a 15th century Italian traveler to the Far East, had apparently been to the Malabar Coast (both wrote memoirs about their travels), and when the Portuguese first landed there with their ships, they encountered a few European merchants, including a Polish Jew who had converted Christianity. However, apparently no ships from Europe had been to the Malabar Coast since the Romans.
6. Wills, J. E. Jr. 2005. *Pepper, Guns, and Parleys*. Figueroa Press.
7. See Note 1 for books outlining such theories at length.
8. Sherman, P. W. and Billing, J. 1999. Darwinian Gastronomy: why we use spices. *BioScience* 49: 453–463.
9. Sherman, P. W. and Hash, G. A. 2001. Why vegetable recipes are not very spicy. *Evolution and Human Behavior* 22: 147–163.
10. Pino, J., et al. 1990. Chemical and sensory properties of black pepper oil (*Piper nigrum* L.). *Die Nahrung* 34: 555–560; Meghwal, M. and Goswami, T. K. 2012. Chemical composition, nutritional, medicinal and functional properties of black pepper: A review. *Open Access Scientific Reports* 1: 2 (doi:10.4172/scientificreports.172).
11. Chaieb, K., et al. 2007. The chemical composition and biological activity of clove essential oil, *Eugenia caryophyllata* (*Syzigium aromaticum* L. Myrtaceae): A short review. *Phytotherapy Research* 21: 501–506.
12. Padmakumari, K.P., Sasidharan, I, and Streekumar. M. M. 2011. Composition and antioxidant activity of essential oil of pimento (*Pimenta dioica* (L) Merr.) from Jamaica. *Natural Product Research* 25: 152–160.
13. Maya, K.M., Zachariah, T. J., and Krishnamoorthy, B. 2004. Chemical composition of essential oil of nutmeg (*Myristica fragrans* Houtt.) accessions. *Journal of Spices and Aromatic Crops* 13: 135–139.
14. Lin, Y.Q., Kong, D.X., and Wu, H. 2013. Analysis and evaluation of essential oil components of cinnamon bark using GC-MS and FTIR spectroscopy. *Industrial Crops and Production* 41: 269–278.
15. A phenyl ring, sometimes also called a benzene ring, is a structure of six carbon atoms linked to each other to form a ring, and in which the carbons "share" their electrons. Such a ring is typically depicted in the following way:

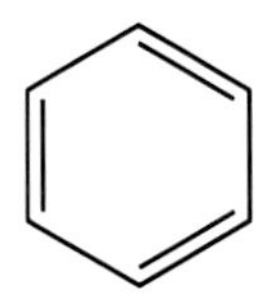

Many organic molecules that have a phenyl ring as part of their structure are odorous to the human nose or to other animals (there are species-specific differences in the ability to smell certain chemicals), so chemists have called such compounds "aromatic." However, not all such compounds are odorous, so in chemistry "aromatic" means having a phenyl ring, but not necessarily being odorous.

16. Muchtaridi, Subarnas, A., Apriyantono, A., and Mustarichie, R. 2010. Identification of compounds in the essential oils of nutmeg seeds (*Myristica fragrans* Houtt.) that inhibit locomotor activity in mice. *International Journal of Molecular Sciences* 11: 4771–4781.
17. This is why hydrophobic compounds are also called lipophilic, or "oil-loving."
18. Szallasi, A. 2005. Piperine: researchers discover new flavor in an ancient spice. *Trends in Pharmacological Sciences* 26: 437–439.

19. Yang, F., et al. 2015. Structural mechanism underlying capsaicin binding and activation of the TRPV1 ion channel. *Nature Chemical Biology* 11: 518–524.
20. Onion and garlic are also heavily used, but typically as fresh plant material and are therefore considered herbs, not spices.
21. "Botulus" is the Latin word for sausage.
22. For a detailed description of the human olfactory system, see Herz, R. 2007. *The Scent of Desire: Discovering Our Enigmatic Sense of Smell*. Harper Perennial.
23. McGann, J. P. 2017. Poor human olfaction is a 19th-century myth. *Science* 356: 597. This article points out that while it is commonly believed that humans have a poor sense of smell compared to other animals, this belief is not based on empirical data and instead relies on a hypothesis of the famous 19th-century neuroanatomist Paul Broca, who posited that the evolution of human free will, which he claimed resided in an enlarged human frontal lobe, required a concomitant decrease in the size of the olfactory bulb. However, recent work has shown that the size of the human olfactory bulb is quite large and it contains a similar number of neurons to that found in the olfactory bulbs of other mammals.

5 Caffeine, Opium, and Other Drugs for the Masses

WHY PLANTS MAKE PSYCHOACTIVE COMPOUNDS

As we saw in Chapter 3, it has been recognized for millennia that the consumption of nicotine affects human behavior. Indeed, tobacco products have always been used by people mainly for the purpose of achieving such psychoactive effects. Some believe that sucrose, also discussed in Chapter 3, similarly affects the behavior of human beings, particularly immature ones, although this seems to be a rather recent notion. But nicotine and sugar are not unique. There is a long list of chemicals produced by a variety of plants that exert some influence on human behavior, an influence that is not caused indirectly by their contribution to metabolism but rather by their direct interaction with the nervous system. These include alcohol, caffeine, theobromine, cocaine, morphine, tetrahydrocannabinol (THC), ephedrine, mescaline, cathinone, some of the spice chemicals discussed in the previous chapter such as myristicin, and many others.

There is also a long history of restrictions on the consumption of such compounds imposed by various authorities, although it appears that the long lists of outlawed substances that many modern states have accumulated in their law books is, well, a modern phenomenon. On the other hand, states and other authorities also have a long history of encouraging the use of some of these substances for various reasons that will be touched on below. In some cases, as with nicotine and alcohol, some authorities in the past appeared (and maybe still do) to both encourage and discourage consumption, torn as they were by eliminating the ill effects of these drugs while wanting to financially benefit from the taxes levied on them, and perhaps too from some of the reputed positive social benefits of their use.

With few exceptions, the common natural[1] psychoactive chemicals (defined as those affecting the brain and therefore affecting behavior) that humans consume are of plant origin. Authoritarian interference with the ability of individuals to consume their preferred psychoactive chemicals and with the ability of producers and suppliers of such chemicals to make a profit always leads to strife. When this strife occurs within a socially cohesive group such as a state, the violence may or may not be contained. But often, the sources and the consumers of such chemicals do not reside in the same country, and in such a situation violence is much more likely to erupt. In this chapter we will describe some of the most common plant-derived psychoactive materials and how the struggles over their use has caused major disturbances to national and international peace. As with previous chapters, the emphasis will be not on the detailed description of their neurological effects on humans, but on the biology of the plants themselves and how this played a role in the violent social events that ensued.

We have already seen that plants make many compounds to defend themselves against bacteria, fungi, insects, and vertebrate animals. These chemicals are "bad" for such organisms in general, but their badness, or toxicity, can come in many different ways.[2] Some can simply incapacitate or kill the organism if they are absorbed in high enough quantities, without drastically changing the behavior of the organism. Others can be both lethal and psychoactive (depending on the dose), and others are mostly psychoactive, although their effect may lead indirectly to the death of the organism.

Humans, like other bilaterally symmetric animals, have a central nervous system (the brain and spinal cord) and a peripheral nervous system.[3] The nerve cells on the periphery – embedded in the body – detect signals via protein receptors found on their outer cell membranes. As already discussed in Chapters 3 and 4, such receptors have affinity to specific "ligands," small molecules that are released by adjacent cells or that come from outside the body (the rod and cone cells in the eye are an exception, having internal receptors for light, not chemicals). Once the receptors bind a ligand, a cascade of biochemical reaction is initiated that leads to the neuron cell "firing," a term that denotes the generation of an electric signal that is propagated along the length of nerve cell to the next nerve cell, which may still be located in the peripheral nervous system or in the central nervous system. Communication between different nerve cells in both the peripheral and central nervous systems occurs via synapses, or junctions. The synapses of neuron cells also contain various receptors that bind specific chemicals. Such chemicals are released by the neuron cell that just fired, and are called "neurotransmitters." The neurotransmitters diffuse across the synapses and bind to the receptors of the nerve cell on the other side of the junction. Once a critical number of such chemicals are released and a sufficient number of receptor-ligand complexes are formed, the nerve cell now fires, generates the same type of electric signal that runs along the length of the cell and arrives at the other end of it, causing a release of neurotransmitters to propagate the signal further. Eventually, the signal arrives in the brain, where it is decoded into a specific "sensation".

Plant chemicals that can change the behavior of humans are those that are capable of binding to neuroreceptors of either the central or peripheral nervous systems, or both. These chemicals may be identical to the ones produces by the human body, or they may be similar yet not identical. We have already seen in Chapter 4 that capsaicin and piperine are able to bind to the TRPV1 receptor of nerve cells whose function is to detect heat. But these plant chemicals are not the neurotransmitters that the TRPV1 receptor usually binds – this receptor binds the compounds anandamide and N-arachidonoyl-dopamine, which are neurotransmitters that the human body makes. So capsaicin and piperine are *mimics* of neurotransmitters, and as we will shortly see, many other plant psychoactive compounds are also mimics of neurotransmitters, at least in the sense that they are able to bind to specific neuroreceptors. However, they may not have exactly the same chemical properties as the endogenous neurotransmitters, and this is why their effect on human perception and behavior may be somewhat different from the effect of the endogenous neurotransmitters. For example, if they can bind to the receptor but their binding does not elicit neuron activation, they are considered antagonists, because, by binding themselves to the receptor, they physically exclude the binding of the natural neurotransmitter and thus neutralize the neuron. On the other hand, if their binding does activate the neuron then they are considered agonists. However, the neurological response of agonists may still be different than one obtained with an endogenous neurotransmitter because they may have stronger or weaker binding affinity to the receptor, or because they may not be degraded as fast as a normal neurotransmitter (neurotransmitters are usually quickly degraded so that they deliver a one-time message when necessary and do not stick around to continuously activate the nerve cell). Yet other plant psychoactive compounds exert their effects indirectly, for example by inhibiting enzymes that synthesize or degrade neurotransmitters.

It appears that natural selection has led to the evolution of plants that can synthesize compounds that mimic neurotransmitters or otherwise interfere with the nervous system. Although it is not always clear what advantages accrue to the plants by making such compounds, several hypotheses have been proposed. First, it should be remembered that many of these compounds may, at high enough concentrations, simply overwhelm the nervous system so as to incapacitate the animal. An immediate cessation of herbivory because of incapacitation is clearly beneficial to the plant. An unpleasant sensation such as the pain caused by capsaicin is also in the plant's interest because it may lead to an interruption in the activity injurious to the plants as well as inducing long-term avoidance of the plant and its congeners by the offending animal, which learns to associate eating the plant with the unpleasant experience.

Before we discuss several of the main plant psychoactive compounds and their contributions to violence among people, a few words about alcohol. The manufacturing and consumption of ethanol (the specific chemical name for the type of alcohol found in wine and most other alcoholic drinks) are very ancient human activities found in almost all societies. Ethanol is an addictive psychoactive drug – it acts as a general depressant and it disables many mental processes – and both its direct use and fights over its use have led to much death and destruction, too numerous to describe here. However, while living plants do make small amounts of ethanol, it does not accumulate in high concentrations in the bodies of living plants. Therefore, in order to make sufficient amounts of ethanol for drinking, human have used plant material – for example, starchy grain seeds, sugarcane juice, or ripe sweet fruits – simply as the carbohydrate sources to feed microorganisms such as yeast that break the sugars down to ethanol in a fermentation process.[4]

Because all major crop plants that are used for making alcoholic drinks can also be used as food, and often are, any bans – and subsequent fights – over alcohol production occur at the manufacturing stage and not at the cultivation stage. For example, grape cultivation is not outlawed in Muslim countries where alcohol consumption is banned, nor was it outlawed during the Prohibition Period from 1920–1933 in the United States. Because the carbohydrates that are the basis for the fermentation process for ethanol production are common plant metabolites that are found in plants growing everywhere, practically all ancient human societies learned to produce ethanol and to consume it, using a variety of plant sources. Today, ethanol is one of the most prevalent psychoactive drugs in use by people everywhere, even in societies where it is prohibited for religious and other social reasons.

CAFFEINE

Another psychoactive drug that is widespread today and that faces almost no legal restrictions is caffeine (Figure 5.1). Caffeine is a chemical that is biosynthesized by plants and only by plants, but not by all plants. Indeed, plant species that make caffeine typically grow in warm climates. Furthermore, prior to domestication, plants containing caffeine originally grew in only a few localities around the world. For this reason, the spread of the use of this chemical and its evolution into a worldwide recreational drug is a relatively recent phenomenon.

Caffeine[5] is a small molecule that contains some nitrogen atoms arranged in "rings" of atoms and is therefore considered an "alkaloid."[6,7,8] It is a toxic compound with an LD_{50} value of 125 mg/kg for mice, and a similar value is estimated for humans (this translates into roughly a third of an ounce for a person weighing 165 pounds). Caffeine is an inhibitor of several enzymes, particularly of the class called hydrolases, that break down compounds into smaller components, thus negatively impacting some metabolic processes. When insects munch on plants containing caffeine, their growth is reduced and they suffer decreased fertility and even death. Long-term consumption of high amounts of caffeine (roughly the amounts present in >10 cups of coffee) by people could lead to hardening of blood vessels and damage to muscle tissue.

Caffeine also exerts a strong effect on the nervous system, which is the main reason for its widespread use by people. Caffeine is an antagonist of adenosine neuroreceptors and several other neuroreceptors. When consumed in moderation, its general effect is to make the person more alert and less tired. It also enhances memory, and not only in humans. The floral nectar of several plant species, such as the coffee and citrus trees, contain caffeine. Bees that visit the flowers drink the nectar, which is produced in nectar glands at the base of the flowers and contains sugars and amino acids as well. The nectar is the reason the bees visit the flowers, as with moth pollination of tobacco flowers described in Chapter 3, and just as happens with the moths, in the process of moving around the flower to drink the nectar the bees pick up some

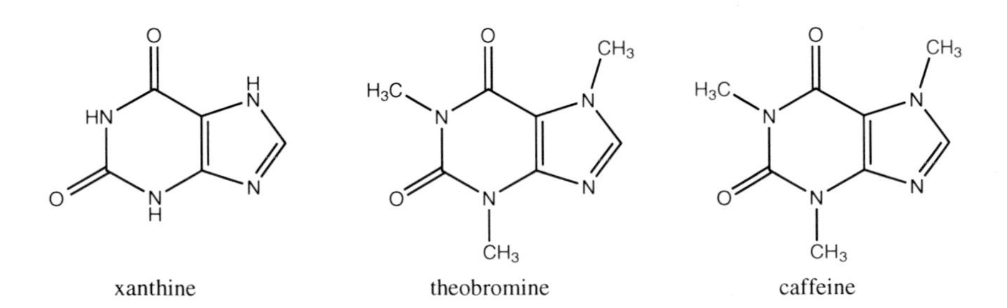

FIGURE 5.1 Structures of the caffeine molecule and its biosynthetic precursors. The methyl groups present in caffeine and theobromine, but not in their precursor xanthine, are shown in red.

pollen grains on their bodies, and some of these pollen grains will deposit on the stigma of the next flower the bees visit, thereby bringing about pollination. Most bees are generalists, meaning that they visit the flowers of more than one species. In control experiments, it was shown that when the bees visit flowers that contain nectar with low levels of caffeine (high levels will make the nectar bitter), they are three times as likely to remember the scent of these flower than the scent of non-caffeinated flowers and to associate these flowers with the rewarding nectar, and therefore more likely to visit caffeinated flowers again. This clearly improves the pollination efficiency of plants that caffeinate their floral nectar.

But due to its toxicity, the main role for caffeine in plants is in defense. The seed-bearing fruit of the small coffee trees (or bushes), *Coffea arabica* and *C. canephora* – the berries from which the coffee drink is made – are one example (Figure 5.2). Cola (*Cola acuminate*) nut and guarana (*Paullinia cupana*) fruits also contain high levels of caffeine. But the presence of caffeine is not limited to fruit and seeds. Plants in the *Coffea* genus also have high levels of caffeine in young leaves and shoots, as do the corresponding anatomical structures of several other plants such as *Camellia sinensis* (the tea plant) (Figure 5.2) and *Ilex paraguariensis* (the South American plant called yerba mate).

The list of plant species that are known to synthesize caffeine for defense or for training their pollinators is eclectic. They are not particularly closely related to each other, and they have many close relatives that do not make caffeine. The most likely conclusion from this observation is that the ability to make caffeine evolved in different plant lineages independently several times,[9] a phenomenon known as convergent evolution. When we look at the chemical structure of caffeine, we can see why it has proven relatively easy for the evolution of caffeine biosynthesis to occur multiple

FIGURE 5.2 Coffee, tea, and cacao plants (counterclockwise from top left).

times in plants. The caffeine molecule looks very much like the compound xanthine, except that the caffeine molecule, but not the xanthine molecule, has three methyl groups – each methyl group consists of a carbon atom bound to three hydrogen atoms – attached to three of the nitrogen atoms (Figure 5.1). Xanthine is a compound found in practically all living organisms, because it is both a precursor for the synthesis of DNA and also a degradation product obtained when organisms turn over their DNA as part of the regular maintenance process. Living organisms also have multiple enzymes that add a methyl group to a nitrogen atom that is part of a larger molecule. Such enzymes are called methyltransferases. It seems that in the past, plant methyltransferases whose original functions was to put methyl groups on some compounds other than caffeine mutated, multiple times in different plant lineages, so that the mutated enzymes acquired – by chance – the ability to bind to xanthine and to add up to three methyl groups to it.

When a plant has acquired three new methyltransferases that each can add a methyl group to xanthine to a different nitrogen atom in the molecule, it thereby evolved the ability to make caffeine (for some unknown reason, even though xanthine has four nitrogen atoms, a molecule that consists of xanthine with four methyl groups has never been discovered in plants). If only two new methyltransferases evolve to add two methyl groups, the result may be theobromine,[10] a chemical with somewhat similar properties to those of caffeine that is found in high concentrations in the fruit of the South American plant cacao (*Theobroma cacao*) (Figure 5.1). If the ability to make caffeine or theobromine was advantageous to the plant, plants with this evolutionary novelty were more fecund, so that they passed this ability to the next generation at a higher rate than other individuals of their generation did.

Throughout history, people who lived close to the land – hunter-gatherers, subsistence farmers, herders – developed extensive knowledge of their local flora and its chemistry. People were always tasting the wild plants they encountered and observing the effects on themselves, as well as noticing the health consequences to their domesticated animals when they grazed on specific plants. It is therefore likely that plants containing caffeine, whose effect on humans is quite pronounced, were discovered early on (interestingly, all plant species that produce high levels of caffeine are native to tropical or subtropical regions of the world). In South America, native mate and guarana plants were consumed as drinks before Europeans arrived in the 16th century, and probably much earlier. Historical records indicate that the tea plant *Camellia sinsensis*, which originated in Southeast Asia, was used to make a herbal infusion at least two thousand years ago, and from China it was spread by people to neighboring countries, reaching Japan by the 7th century. The two tree species from which berries are collected to make coffee – *Coffea arabica* and *C. canephora* – are native to Ethiopia, with the latter species having a somewhat wider native distribution that includes also Western and Central Africa. Historical mention of coffee was made in Arabic medicinal books by the 9th century, and by the 15th century coffee cultivation moved outside this region to Yemen, across the Red Sea.

The introduction and use of coffee and tea in Europe, a continent in the temperate zone that lacks plants containing caffeine, initially followed somewhat different trajectories. Coffee was actually introduced to Europe before tea. As the Turks established their empire in the Middle East at the beginning of the 16th century, they were introduced to coffee from the Arabian Peninsula, and coffee house became common throughout the Ottoman Empire. Soon, coffee was traded between the Turks and Europe mostly via Mediterranean shipping lines.

On the other hand, while Arab traders operating on the Silk Road were certainly familiar with tea drinking in China, tea was not one of the trade items that Arabs regularly brought back from China. The Portuguese encountered tea when they arrived in China and Japan in the first half of the 16th century, but they too did not think much of it and tea remained a local drink of the Far East until the Dutch West India Company (VOC) began importing it to Europe at the beginning of the 17th century. The Dutch eventually had to compete with the British East India Company (EIC), which began importing tea to England in 1644.

At the beginning of the 17th century, Dutch ships visited Yemen ports on the way back from the Far East and brought some coffee beans to Europe, but the VOC's trade in coffee was slow to gain momentum. In 1688, VOC began coffee plantations in Java, where the coffee trees thrived, and soon Dutch import of coffee to Europe grew substantially. Once tea and coffee became highly desired commodities in Europe, various European empires planted tea and coffee plants in their tropical and subtropical colonies around the world, with tea cultivation outside China taking longer to establish (more on this to follow).

In a short time after their introduction as luxury items, coffee and tea became very popular drinks in Europe. The two drinks have very different flavors, which are due both to the presence of distinct chemicals besides caffeine, many of them defense compounds, that are found in one of them but not the other, as well as to chemicals that are produced by the breakdown of the natural chemicals in the post-harvest processing. Such steps may involve, for both staples, drying and roasting. However, it was realized from the very beginning, without any knowledge at the time that each contained caffeine, that both drinks produced a similar effect on the mental state of the drinker – alertness and relief of fatigue. Interestingly, it was also noted that tobacco smoking produced a somewhat similar yet distinct state of mind to caffeine consumption, and that simultaneously smoking tobacco and consuming caffeine created an additive effect, boosting mental alertness and well-being even higher. This observation has been confirmed by modern science, which has shown that caffeine and nicotine bind to different sets of neuroreceptors.

Initially, coffee was cheaper in Europe than tea, mostly because of lower transport costs as the source of coffee was closer to Europe than the source of tea. But both coffee and tea were still expensive and therefore their use was first limited to the upper classes. As their prices began to decrease with increasing import volumes, their use spread throughout society (chocolate, the drink made from the theobromine- and caffeine-containing cacao beans, remained a drink of the nobility for much longer). Because the active ingredient in tea and coffee, caffeine, is the same, people have always tended to drink mostly one or the other of these drinks, but not both. In fact, this division was often seen at the country level. For example, England initially saw more people drinking coffee, but in the 18th century there was a general change that social scientists are still at a loss to explain and England became a nation of tea, rather than coffee, drinkers.

Until tea and coffee drinks became common in Europe, people there consumed quite a bit of alcohol – for breakfast, lunch, and dinner – as water was generally unsafe to drink. Alcohol is a depressant, while caffeine makes people more alert and less tired. Since consumption of nicotine, which increases a person's alertness and stamina, was also on the rise in Europe at the same time that caffeine consumption became common, it is not surprising that some social scientists have hypothesized that nicotine and caffeine were important contributors to the general pickup in the economic development of Europe that followed the introduction of these chemicals. These benefits accrued not only to economic activities: tobacco smoking as well as caffeine consumption were soon picked up by soldiers, helping to invigorate militaries and improve their abilities to carry out violent activities. In fact, caffeine-containing (and sugar-containing) foods as well as tobacco were soon included in military rations, and they continue to be included in modern rations, although the recognition of the health hazard of tobacco has recently led some militaries, such as that of the United States, to drop tobacco from their rations (but they can still be purchased on military bases).

In the late 18th century in England, as the Industrial Revolution was picking up steam, tea consumption was spreading to the lower classes. British cotton mills were employing children as young as five years old and young women. The workday could be up to 16 hours long, and food, which was provided by the factories, was meager and of poor quality. The caffeine in the tea could help keep these overworked and underfed workers from falling asleep on the job, falling into the machinery and causing major monetary damage to the mill owner (the workers were often seriously hurt and even killed in such accidents). Tea, while still more expensive than beer (but cheaper than coffee, when calculated on the price per cup), was therefore occasionally included in the drink provided to the workers. Workers sometimes bought tea on their own, and there were in fact outfits in England that recycled used tea leaves and sold them at a reduced price. Also, the British began to add milk and or sugar to their tea. As sugar became cheap too (see Chapter 3), sweetened tea became the perfect junk drink[11] for the overworked and underpaid worker, making a significant contribution to the total calories consumed and keeping the worker awake.

The trading of tea and tobacco played a central role in the rebellion of the English colonies in North America against English rule. Short on revenue at the end of the Seven Year War in 1763, English Parliament looked for ways to raise revenues from the colonies, with the justification of using the money to protect the colonies themselves. Between 1764 and 1773, parliament passed several acts (and later repealed some of them) that, among other things, increased levies on colonial tobacco farmers and further restricted their ability to sell their products to non-English merchants, and also taxed several commodities imported into the colonies, including tea. The Tea Act of 1773 was the last straw. The purpose of this act was actually not to directly increase tax but to help the British East India Company (EIC) get rid of its large unsold inventories of tea in England. Until then, the EIC did not sell tea directly to the colonies but to merchants in England who then exported it to the colonies, paying a tax (instituted by the Townshend Acts of 1767) that was used to pay the salaries of colonial officials. However, tea smuggled into the colonies by the Dutch was cheaper because tea imported into Holland was not taxed by the Dutch government, while tea imported into England was taxed, and consequently 60% of the tea sold in the colonies came from smuggling. The Tea Act of 1773 allowed the EIC to sell tea directly in the colonies and eliminated the tax on the tea that the EIC was required to pay in England, but the EIC was still required to collect the three pence per pound tax, stipulated by the Townshend Acts, from the American merchants. The Tea Act elicited strong resistance from the colonists for reasons that for brevity's sake will not be detailed here, and it would suffice to say that it came after a long series of events that launched the American patriot movement toward a principled opposition to English rule. At any rate, the patriots were able to pressure most American merchants selected by the EIC to refuse to accept the shipments of tea, but when they did not prevail with the Boston merchants they staged a raid on the EIC ships and dumped the tea cargo into the water, an event known as the Boston Tea Party. The American Revolution was then well on its way.

OPIUM AND THE OPIUM WARS

In addition to contributing to the armed conflict with its American colonies in the 18th century, the physiological and economic dependence on tea that the British people developed led them to two infamous military confrontations with China in the 19th century, called the Opium Wars.[12] But before these wars can be described, the second plant contributor to these wars, the opium poppy (Figure 5.3), and the compound present in this plant that is the reason people grow and consume the plant, need to be described.

FIGURE 5.3 Opium poppy, with mature fruit on the right.

There is archeological evidence from European sites for human use of the opium poppy plant (*Papaver somniferum*) going back 6,000 years. The plant was cultivated by Sumerians in Mesopotamia by 3400 BC, and it was well known to the ancient Egyptian, Phoenician, Greek and Roman cultures. The plant spread eastward to India with Alexander the Great. After the fall of the Western Roman Empire, knowledge of the plant was eventually lost in Europe. The plant continued to be cultivated in Muslim countries and from there it spread to China, where it is first mentioned in a document dated to 1483. It was reintroduced to Europe in the 16th century from the Ottoman Empire.

Opium poppy plants produce a mixture of "opiate" alkaloids that include morphine, codeine, and thebaine (Figure 5.4), three compounds that differ by the number of methyl groups present.[13] All three have a methyl group attached to the nitrogen atom of the molecule, but codeine has a second methyl group attached to one of the oxygen atoms and thebaine has a third methyl group attached to another oxygen atom. These compounds are present everywhere in the aerial parts of the plant, but the fruit ("seed pod") is particularly rich in them. When the seed pod is scored with a knife, it exudes a white liquid called latex that contains up to 12% morphine, somewhat less codeine, and substantially less thebaine (the related species *P. orientale* and *P. bracteatum* make high levels of thebaine). Enzymes inside special cells in the plant synthesize the opiate alkaloids, using common amino acids as the starting material, but involving a long and complex set of enzymatic reactions that evidently evolved only in this genus of plants. The final opiate products are then stored in special cells, called laticifers, which are elongated cells that often form a network of their own and may also be located next to the vascular tissue (i.e., "veins") of the plant. The laticifers produce other compounds that allow the opiate alkaloids, which are not very soluble in pure water, to stay suspended in solution, giving the liquid in the laticifers – the latex – its white color.[14] When the plant organs are cut or pressed, the contents of the laticifer cells come out to the surface directly or through ruptured veins.

Since ancient times, the dry latex, called opium, was used to treat a variety of ailments, including diarrhea[15] and coughing, by dissolving it into a beverage, usually containing some alcohol to make it more miscible, and drinking it. Opium is also a sedative, and its superb ability to relieve pain was known to the Greeks, Romans, and other ancient cultures. Drinking a solution of opium imparts pleasant, even euphoric, feelings, so its recreational use also dates to ancient times. When morphine was purified from the opium paste in 1803 by the German chemist Sertürner, studies with this

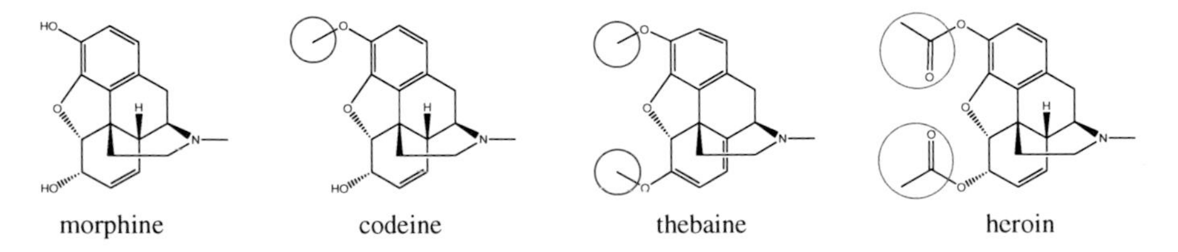

FIGURE 5.4 Structures of the natural opiate alkaloids morphine, codeine, and thebaine, and the synthetic opiate alkaloid heroin. The three natural opioids differ only in the number of methyl groups attached to oxygens (circled in red: none in morphine, one in codeine, two in thebaine). The synthetic opiate heroin has two acetyl groups in these positions (circled in blue) instead of methyl groups.

pure chemical showed that it was the compound mostly responsible for the medicinal and psychoactive properties of opium.

More recently, morphine was shown to act as an agonist of the opioid neuroreceptors that are found both in the central and peripheral nervous systems – it mimics the endogenous neurotransmitters of the receptors, which are chemicals called endorphins that are made of several amino acids linked together. Overconsumption of morphine can lead to death by asphyxiation because the area in the brain that controls lung function stops sending the signal to the lung muscles to contract so breathing ceases, providing one explanation for the defensive value for the plant in producing this compound. Morphine is also an extremely addictive drug. Codeine and thebaine have somewhat overlapping effects with morphine, although in general they are not as potent as morphine. On the other hand, heroin, the semisynthetic opiate that was first made in Germany in 1874 by chemically adding two acetyl groups to morphine, is more than twice as effective as morphine in its analgesic and euphoric activities and is also faster acting because it more efficiently crosses the blood–brain barrier.

After opium was reintroduced to Europe as a medicine, typically called laudanum, in the 16th century, its use became widespread. In fact, it was common for mothers to give it to babies, to treat diarrhea but also simply to calm nervous and crying infants. There is evidence that it was often given to babies of poor women to keep them asleep so that their mothers or caregivers could carry on their work. Use by adults was also common. And as European maritime commerce with the Far East expanded, opium from India began to compete with Turkish opium, particularly in England after the EIC established a strong base in India.

Absorption of morphine through the intestinal system is slow, and much of it is degraded by enzymes in the liver before it reaches the central nervous system. Thus, the effects of opium when ingested are somewhat muted. Today, morphine can be injected with a syringe directly into the bloodstream, causing a very quick response, but this is a recent development. In the 17th century in China, another efficient form of morphine delivery to the central nervous system arose, and that was by smoking opium. Smoking tobacco was introduced to China by the Portuguese in the 16th century. Soon, opium, which was already cultivated in China, was mixed with tobacco, although it is not clear exactly how this innovation originated. In the waning days of the Ming Dynasty, in the first half of the 17th century, the Imperial Court declared a prohibition against smoking tobacco, including tobacco mixed with opium, but smoking pure opium was exempt, as it was considered a medicine (the ban was lifted when the Qing Dynasty took over in 1644). In the technique that eventually developed for smoking opium, this substance was warmed up by a flame so that it began to evaporate but not burn, and the smoker drew the vapors into the lungs using a long pipe. Absorption of morphine into the blood vessels in the lungs had an immediate impact on the nervous system that was much stronger than drinking laudanum, causing a euphoric rush and mental incapacitation. Smoking opium also led to much higher rates of long-term addictions than drinking it.

Opium smoking was first a habit of the upper classes in China because of its cost. Cultivation of opium poppies in China was highly restricted by the government and therefore domestic supply was quite limited. The first foreign traders to see a commercial opportunity in importing opium to China were the Portuguese and the Dutch. They bought opium produced in India and sold it in Canton (present-day Guangzhou), which was practically the only city in China that the Qing emperors allowed to trade with foreigners. By the mid-18th century the British EIC was in control of the Indian subcontinent and the opium trade in India, and it was therefore the British who benefitted the most from selling opium to China, although opium produced in Turkey also found its way to China, with the involvement of, among others, American ships and traders. The increase of opium import into China lowered its price so its use spread, although it was still concentrated around the coast and the big cities inland, with the goods disseminated from Canton by Chinese merchants.

The ill effects of opium smoking – temporary physical incapacitation, addiction, and the social and economic damage to the individual and the country that ensued – were noticed early on as the habit spread in China. An Imperial ban on selling opium, but not on using it, was promulgated in 1729. It was followed in 1799 by a much stricter ban on importers, local sellers, and users, who were all threatened with severe punishment, but enforcement was lax and easily circumvented by bribes. To legally comply with the ban on import, the EIC developed a strategy of selling the opium in India to other British merchants, with the stipulation that the opium could only be resold in China. These independent merchants then shipped it to China and smuggled it into the country by conniving with local officials.

Until 1833, the EIC had a monopoly on trade between England and the Far East. The main item of trade that the EIC bought in China to sell in England was tea. There was little interest in China in Western goods (except opium, which was illegal) and Chinese tea merchants wanted to be paid in bullion silver, the metal that served as the de facto currency in China but which was in short supply in England. The EIC trade in Indian opium was designed to solve this problem. The independent British merchants who sold the opium in China were paid with silver, and they used their silver to buy the opium from the EIC in India. In turn, the EIC used this silver to buy tea in Canton and ship it to England.

This situation worked for a while to the benefit of everyone involved in the trade, although not necessarily to the customers in either China or England. Both the British and the Chinese traders made fortunes,[16] the British authorities collected vast sums in taxes on the opium trade in India and the importation of tea into England, and Chinese authorities benefitted from taxing the Canton merchants (but not opium directly, since it was illegal) who passed the cost to the customers. But as the quantities of opium imports kept increasing and its price in China kept going down, the number of users – and addicts – increased, which led to more opium imports, creating a vicious cycle. By 1820, the overall net positive flow of silver into China, which came mostly from the Chinese trade with the Spanish in the Philippines,[17] reversed course, because the Spanish trade was waning (the silver mines in South America were exhausted) and China was now paying more in silver for opium and a few other items of trade from the EIC than they were getting for the goods they sold the company.

Then, with the era of mercantilism in the West coming to an end because of the rise of the new ideology of free trade, in 1833 the British parliament revoked the monopoly of the EIC on trade with China so that other British companies could participate and benefit from this trade and the British and Chinese subjects could benefit from lower prices due to competition. For once, economic theory worked as advertised. With additional companies wanting to do business in China and needing to import opium to obtain the silver for buying the Chinese goods, opium prices went down, and the number of addicts ballooned. It was estimated that in 1838, on the eve of the First Opium War, there were between 4 and 12 million opium addicts in China out of a total population of approximately 400 million.

With the huge increase in the number of opium addicts and the consequent impact on the economy, the Chinese authorities became very alarmed and tried drastic measures to stop the importation of opium. A special government official, Lin Zexu, was sent to Canton, and he began to enforce the ban on smoking opium, meting out severe punishments to users. Lin told the foreign traders that smuggling opium into Canton would no longer be tolerated, and he next confiscated 1,400 tons of opium stored in the foreign companies' warehouses in Canton (mostly belonging to British traders) and on the ships in open water and proceeded to burn all of it. In response, England began a war, now known as the First Opium War, to force China to allow opium imports into the country. The purpose of the war was, England claimed, to enforce a universal principle – free trade – and, once enforced, it would become evident even to the Chinese that it is in China's best interest to adopt this principle.

Of course, the situation was more complicated. Trade between European countries and China had been fraught with difficulties from the very beginning, when the first Portuguese ship arrived there in 1513. The Europeans were used to a multistate system such as Europe, where different sovereign states traded with each other on equal footing. Outstanding issues, including trade relationships, between European states were discussed by ambassadors that countries exchanged, and these ambassadors had diplomatic immunity. This system was completely unrecognized in China. China and its emperors considered China to be a supreme state above all others, and the only relationship that they recognized with other countries was one of vassal states subordinate to China. They did not exchange ambassadors – they received tribute bearers from other countries. Such tribute bearers from each country came to Peking every few years (there was a schedule dictated by the Chinese bureaucracy), brought gifts with them, kowtowed in front of the Emperor, and could then make requests. There were no negotiations and the Emperor, with the help of his ministers, then decided whether to accede to the requests.

From the beginning, the European countries sought to establish permanent embassies in Peking, and the Chinese would have none of it. The Portuguese and Dutch tried several times, and the British government sent two missions to Peking, in 1793 and again in 1816, with requests to establish permanent embassies, but all were denied. In addition to the prestige issue – China simply did not see herself as equal to any other country, but superior to all – the Chinese state, run by the mandarin class, did not value trade highly. After only a short period in Chinese history in which maritime expeditions and trade extended all the way to Africa, the Chinese emperors forbade sea travel by the middle of the 15th century and closed the country to foreign visitors, believing that China did not need any goods from the outside. While this ban was not absolute – for example, China continued to import pepper and other spices – the Chinese authorities did not want to have much to do with trade issues, leaving Chinese traders to handle the European merchants and limiting access of foreign ships to Canton, at the mouth of the Pearl River. This situation was frustrating to the European traders, since when problems arose they needed to discuss them with the Chinese traders who then talked to the local governor who then contacted the authorities in the capital Peking. The decision of the central authorities in Peking would then be communicated back through these same channels. This process took a long time and often resulted in unsatisfactory outcomes, at least as far as the European traders were concerned.

With the abolition of the China trade monopoly of the EIC, Lord Palmerston, the British Foreign Minister, appointed first Lord Napier in 1834 and then, after his death, Captain Charles Elliot in 1836, to coordinate British commercial activity in China. Both Lord Napier and Captain Elliot had navy ships to back them up, which they used locally, and indecisively, around Canton to try to force the Chinese into changing their policy about not allowing opium importation. The Chinese did not budge, and eventually Elliot was instructed to make a long list of demands to the Chinese authorities. These included a permanent embassy in Peking, the opening of additional Chinese ports for trade, free movement for British ships in inland waterways in China, exemption of Englishmen from the laws of China (i.e., extraterritoriality, meaning that Englishmen in China would be subject to British law adjudicated by British judges), legalization of the opium trade, financial restitution for the damage done to the opium merchants whose opium had been burned, and indemnities for the cost of any military actions that England had to engage in to protect its interests.

The Chinese refused to negotiate, although Lin sent an open letter to Queen Victoria asking her why she would allow her subjects to sell such a vile drug to China when opium was not allowed in England itself. (He was misinformed: opium was legal in England at the time and there were indeed many addicts, but since most users drank it rather than smoked it, the effect was less severe and, furthermore, since most of the addicts came from the lower classes, the English authorities were not too concerned about the problem.) There were certainly many people in England who also thought that British merchants should not be selling a drug in China that was causing such devastation and that was explicitly outlawed there. The response to this argument, by Lord Palmerston and other contemporaries, has a very modern ring to it: the merchants were simply providing a product that the consumers wanted to buy, and if the British companies did not sell it in China (and make a huge profit), companies from other countries would step in to fill this need. If the Chinese government wanted to eliminate opium dependency in their country, the argument continued, they should eliminate the demand, not try to stop the supply.

With the Chinese government refusing to negotiate on the new demands, Elliot commenced military operations against the Chinese in 1839. Additional British ships and troops arrived from India in 1840. The British navy included a modern steamboat and heavy guns, and the British soldiers were equipped with up-to-date rifles. In contrast, the Chinese soldiers were mostly equipped with bows and arrows, spears, and a few extremely ancient muskets. In addition, many of the Chinese soldiers were opium addicts. The outcome of this war was therefore a forgone conclusion – the Chinese armies, when they actually chose to fight and not run away, suffered large casualties, and the victorious British soldiers often carried out civilian massacres, rapes, and widespread looting afterward. Canton fell in 1841, and after the British extended their campaign up the east coast, conquering several port cities, including Shanghai at the mouth of the Yangtze River, and then sending their ship up the river and capturing Nanking (Nanjing) in 1842 (Figure 5.5), the Chinese Emperor was ready to sign a peace treaty with the English. In the Treaty of Nanking signed in 1842, the Chinese acceded to practically the entire set of demands made by the English, and in addition the latter were allowed to keep Hong Kong, which they had occupied during the war. Curiously, however, the British did not bring up the issue of legalizing the opium trade and therefore the subject was completely ignored in the treaty.

The Nanking Treaty also included a provision that gave England a "Most Favored Nation" status, meaning that England was entitled to the best terms of trade with China compared to any other country. Seeing how weak China was, other Western countries, including France and the United States, soon negotiated with China treaties similar to the Nanking Treaty, giving them Most Favored Nation status too.

The Nanking Treaty indeed opened up five ports in China (Canton, Xiamen, Fuzhou, Ningbo and Shanghai) for foreign trade, and while the import of opium was not officially legalized, the Chinese authorities no longer made any efforts to stop it. As a consequence, more opium was being sold in China and more Chinese became addicted to it. But while the opium trade was not impeded, the British had other reasons to complain. For one, the Chinese never fulfilled their obligation to allow a British embassy in Peking. Also, the British wanted to be able to trade anywhere in China, not just in the five treaty ports. Another major issue was a new one – with the waning of slavery, imperialist countries were looking for sources of cheap labor in the form of indentured servants to serve in their colonies (for example, in sugar plantations), and

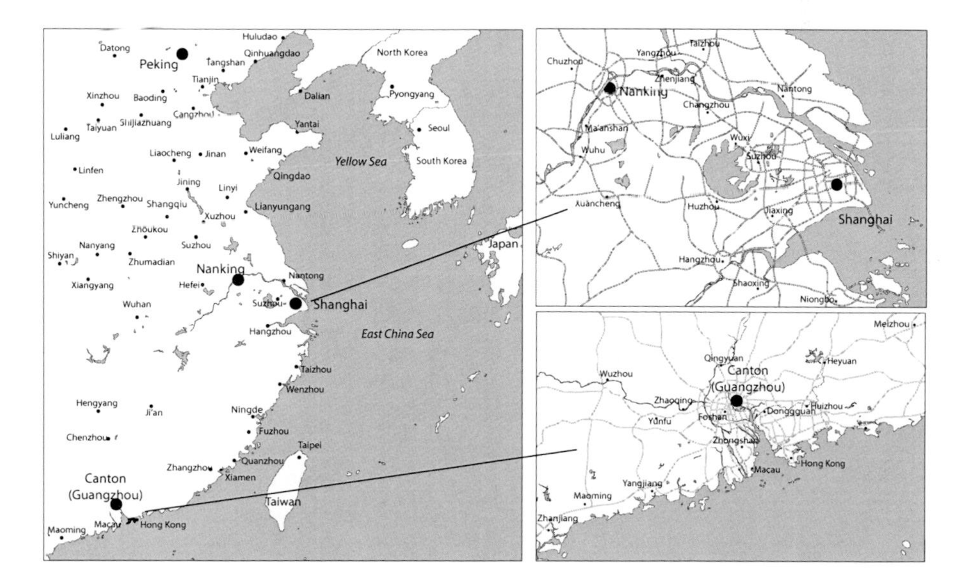

FIGURE 5.5 Eastern China, with the locations of the main battles in the First Opium War shown enlarged on the right.

China had a large number of poor people willing – or forced – to sign up as "coolies." The Chinese government tried to protect poor Chinese from the coolie trade, and this was another "unreasonable trade restriction" that the British wanted to remove.

The opportunity for the British to force the Chinese government into further concessions came in 1856 with the convergence of several events. The Americans and French were renegotiating their treaties with China, as allowed by their respective treaties, and England, as a most favored nation, demanded the right to renegotiate the Treaty of Nanking. China was also in the midst of a civil war, called the Taiping Rebellion, which started in 1850 in the south but eventually spread all the way to Nanking, which the rebels held as their capital. This civil war, which is estimated to have resulted in anywhere from 20 to 70 million casualties by the time the rebels were defeated in 1864, greatly weakened the Qing Dynasty's hold on China. Finally, England used as a pretext the capture by the Chinese authorities in Canton of what the English claimed to be a British-registered ship and its Chinese crew, a violation of the extraterritoriality clause of the Nanking Treaty (it turned out eventually that the British registration of the ship, The Arrow, had expired). It helped fan British public sentiment at home in favor of war when, according to an English eyewitness to the event, the Chinese soldiers pulled down the British flag from the mast of the boat in a most degrading manner.

Hostilities were first localized to the Canton area, but Lord Palmerston, who was by then the prime minister, decided to take this opportunity to press the military advantage to gain more concessions from China, including an embassy in Peking, more ports open to trade, and the legalization of the opium trade. The French, too, saw an opportunity and sent a navy force to join the British campaign, using as their *casus belli* the execution by the Chinese authorities of a French priest who had been living and proselytizing in an inland village out of bounds for foreigners. The Americans, who were against the opium trade, had US Navy gunboats in Chinese water but declared neutrality, while the Russians, who had their own territorial designs on coal-rich northeast China (see Chapter 7), sent no military force but a few observers to join the British expedition.

In 1858, the British and the French forces, under the overall supervision of Palmerston's plenipotentiary to China, the 8th Lord Elgin, attacked and conquered military forts on the coast near Tientsin (present-day Tianjin), only 150 km from Peking. Representatives of the Emperor then quickly signed the four Tientsin Treaties with the English, French, American, and Russian representatives, acceding to most of their respective demands. However, the opium trade again was not discussed, and the Chinese only agreed to small diplomatic legations, not full embassies, in Peking. Hostilities then subsided for a while, with the exception of some lethal guerilla attacks by Chinese forces against British forces in Canton, and retaliation by the British. Lord Elgin returned to England to serve in the government as Postmaster General.

However, when in June 1859 British and French ships arrived again at Tientsin, with American ships not far behind, and then tried to sail up the Bei He River to Peking to set up their diplomatic legations, the Chinese refused them passage, inflicting heavy casualties on the European forces[18] and renouncing the Tientsin Treaties. In response, a year later Lord Palmerston, who was again the prime minister after a brief stint in the opposition, sent Lord Elgin back to China with a force of 11,000 soldiers.

The French also sent additional troops, and the combined British-French expeditionary forces then fought their way all the way to Peking and laid siege to the city, decisively defeating all Chinese forces trying to block their progress. The emperor fled, and his brother, Prince Gong, then ratified the Tientsin Treaties, and, on October 18, 1860, signed the Peking Convention, which, among other additional concessions, finally legalized the import of opium in China, albeit in the indirect way of imposing an official tax on its sale. To conclude the Second Opium War, on the same day that the Peking Convention was signed, Lord Elgin gave the order to destroy the Summer Palace outside the city walls (it was first thoroughly looted)[19] as punishment for the torture and murder of some British and French military personnel and civilians that had been captured by the Chinese during the campaign.

With the legalization of opium in 1860, the number of Chinese addicts kept increasing. In 1879, opium imports, at seven million tons, were double what they had been in 1859. The British government was deriving 15% of its total revenues from taxation on opium, so that even the famous British politician Gladstone, who as a member of parliament had been a fierce opponent of the Opium Wars on moral grounds, now embraced it as prime minister. This left the Chinese no choice but to encourage the local cultivation of opium poppies, leading to a decrease in opium imports starting by the early 1880s. Interestingly, the British had already begun doing the same thing with tea. While a tea variety (*Camellia sinensis var. assamica*) grows naturally in India and adjacent countries, the British had brought Chinese tea (*Camellia sinensis var. sinensis*) seedlings to India and Ceylon (Sri Lanka) in the 1820s, and from the 1850s tea from the Indian subcontinent constituted a large portion of the tea imported to England and its colonies.

The growth in domestic opium cultivation in China, however, made the opium addiction problem even bigger. A large proportion of the public was addicted, including high officials and even members of the Imperial family such as the Empress Dowager Cixi, and this led to economic and political instability and further social strife that sometimes erupted into major violent events. One of these incidents was the Boxer Rebellion in 1899–1901, which was partially directed against foreigners and involved again the armed forces of major European militaries as well as the militaries of the United States and Japan. Although opium was banned again in China in 1906, elimination of widespread opium addiction in China occurred only after the communists took over in 1949 and enacted a total ban on opium poppy cultivation and opium import and use, a ban that was ruthlessly enforced.

COCAINE, TETRAHYDROCANNABINOL, AND MODERN WARS

Opium has continued to contribute to violence in modern times. Globalization – mostly worldwide commerce – has intensified since the Opium Wars, not least so when it comes to trade in psychoactive plant materials. Besides morphine and caffeine, two other notable plant-derived psychoactive compounds that have achieved wide popularity in the modern era are cocaine, an alkaloid synthesized uniquely in the leaves of the South American coca plants (Erythroxylum coca and a related species *C. novogranatense*), and tetrahydrocannabinol (THC), made exclusively in the leaves and floral brackets of the marijuana plants (*Cannabis sativa* and related species) (Figure 5.6). Neither of these two compounds can be made synthetically on a commercial scale, so people have to grow these plants as agricultural crops in order to obtain these chemicals.

The coca plant is native to the eastern slopes of the Andes in South America and has been cultivated there for thousands of years. Although in recent times it has been successfully cultivated in other tropical parts of the world, such as Java, today the bulk of the cocaine traded around the world comes from South America. Cocaine is a stimulant. When the leaves are chewed, the cocaine in them is slowly absorbed into the bloodstream and the effect is to combat hunger, thirst and fatigue, and to increase alertness and well-being. Cocaine is also a strong topical analgesic.[20] However, when cocaine powder or vapor is quickly absorbed through the nose or the lung, it causes a short burst of euphoria, and it is for this effect that it is mostly consumed today outside South America. Cocaine exerts its neurological effects by inhibiting the recycling of neurotransmitters in the brain such as serotonin, norepinephrine, and dopamine, causing these neurotransmitters to linger in the synapses and continue to stimulate the nerve cells to generate signals, thus leading to a highly excited mental state. And it is a highly addictive substance.

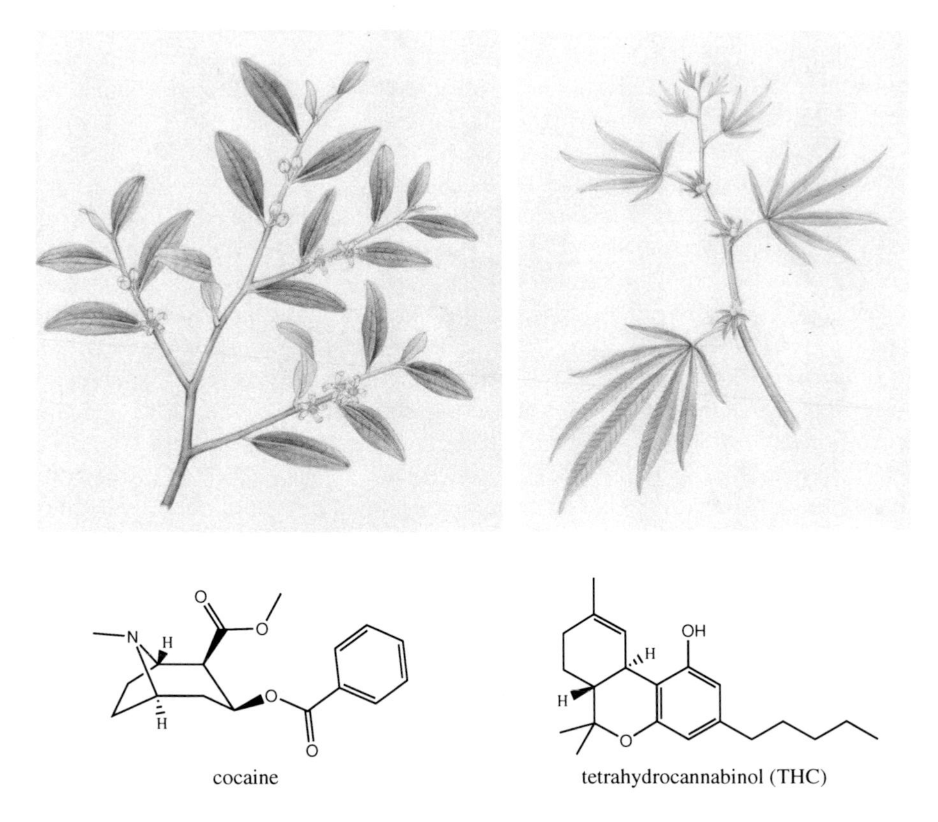

FIGURE 5.6 Left, a branch of the coca plant, *Erythroxylum coca* (top), and the cocaine chemical structure (bottom). Right, a branch of a female *Cannabis sativa* plant (top), and the structure of tetrahydrocannabinol (bottom).

The *Cannabis* genus of plants is native to Central Asia and the Indian Subcontinent. Trichomes on the leaves of the *Cannabis* plants, and particularly on the numerous brackets at the base of the female flowers,[21] synthesize and store high amounts THC as well as related compounds. Numerous scientific experiments have shown that THC is responsible for most of the psychoactive effects that are caused by smoking or ingesting *Cannabis* material. THC is a mild analgesic and a mental relaxant, but is also an appetite stimulator, an effect obtained via binding to the cannabinoid receptors. Herodotus in his *Histories*, written in the 5th century BC, is believed to be describing the use of *Cannabis* plants by the Scythians when he recounts how plants were burned inside a closed tent and the smoke inhaled, since the described psychoactive effects on the inhalers resemble those of THC.

Frequent use of and addiction to opiates and cocaine alkaloids have extremely bad consequences for the health and social and financial well-being of users. THC, which is not an alkaloid (it has no nitrogen), appears to have a much weaker effect on behavior and to cause little or no addiction. But ironically, the greatest contribution of all these psychoactive plants to war and destruction comes from efforts to ban their cultivation and use.[22] Such efforts have intensified since the beginning of the 20th century[23] as modern states have tried to regulate the use of all manners of chemicals, with the stated goal of protecting individuals and society from the bad consequences of such use.

Making it illegal to grow these plants and to extract and sell their psychoactive compounds has made these consumer products, which appear to have high demand regardless of the legal situation, expensive and therefore financially lucrative to growers and sellers. Since marijuana plants are grown almost anywhere in the world, and thus growers are in close proximity to users, revenues from international trade in marijuana, while not inconsequential, are much lower than such revenues obtained from cultivation and export of opiates and cocaine. Indeed, revenues from the opium trade have been funding one or the other side in the constant wars that have plagued Afghanistan since the late 1970s, wars that have cost millions of lives and have devastated the country. In South America, various guerrilla forces have used revenues from cocaine cultivation and trade to finance military operations. Most notable of these was the organization called the Revolutionary Armed Forces of Colombia (FARC is its Spanish acronym), which, until they signed a peace agreement with the government of Colombia in 2016, had carried out a 52-year war against the government that resulted in at least 220,000 deaths and the creation of five million refugees. The high profits and illegal status of these psychoactive plants have also led to much fighting between criminal gangs for control of the trade in these plants and their chemicals in producing, transit, and consumer countries, with many casualties among gangs as well as civilians. In some cases, governments have lost control of major parts of their territories to such gangs, as has happened in several Central American countries.

CONCLUSION

Different lineages of plants have evolved the ability to make many chemicals that have adverse physiological effects on the animals that munch on them. Some of these compounds also have psychoactive effects on animals, presumably because such effects interfere with and limit herbivory. Humans have not been, in most cases, the actual herbivores of these plants in nature, but due to constant observation on the effects of consumption of such plants by livestock and experimentation by humans, people have discovered the psychoactive effects of such plants. Consequently, people have adopted the use of these plants because they feel that the psychoactive chemicals in them enhance their performance or make them happier, at least temporarily. While some people view such hedonistic pursuits as totally positive, in practice consumption of such plant toxins often causes a number of bad consequences, including diminished mental and physical performance, bad health (including addiction), and economic and social impairment.

Such negative effects have led many to conclude that it is not in the long-term interest of individual citizens to consume these products, nor is it in the interest of the society in which they live. Indeed, governments everywhere have regulated or completely banned the growing and consumption of some of these plants. Realistically, however, it appears that despite all these efforts, cultivation and sale of such plants have really taken off in the last two hundred years, and this trend does not seem to be abating. The human population continues to increase, so demand is increasing. And with globalization, when there is demand anywhere in the world, there will be supply somewhere, and someone will make a profit. The unfortunate outcome has been that efforts to stop the use of such plants, sometimes called "Wars on Drugs," have by themselves often resulted in large numbers of casualties that include not only many dead plants but also many dead people.

NOTES

1. While some animals and microorganisms make defense toxins that have psychoactive effects as well, they are not available in large amounts and not widely used. Lysergic acid diethylamide (LSD), on the other hand, has been more popular but it is actually a synthetic compound made by modification of the natural chemical ergotamine, made by the ergot fungus that grows on wheat and related grasses.
2. Harborne, J. B. 1993. *Introduction to Ecological Biochemistry*. 4th ed. Elsevier Academic Press.
3. A good introduction to the neuroscience aspects of the topics discussed in this chapter can be found in Bear, M. F., Connors, B. W., and Paradiso, M. A. 2016. *Neuroscience: Exploring the Brain*. 4th ed. Wolters Kluwer.
4. Stewart, A. 2013. *The Drunken Botanist*. Algonquin Books of Chapel Hill.
5. For a general background on caffeine and human association with it, see Weinberg, B. A. and Bealer, B. K. 2001. *The World of Caffeine*. Routledge.
6. Alkaloids are a class of organic compounds that contain at least one nitrogen atom attached to two or more carbons and arranged as part of a ring of atoms. The presence of the nitrogen makes the compounds basic, or alkaline, so the pH of the solution containing such compounds is >7 (absent other moieties that impart a negative charge).
7. Miyashira, C. H., et al. 2012. Influence of caffeine on the survival of leaf-cutting ants *Atta sexdens rubropilosa* and in vitro growth of their mutualistic fungus. *Pest Management Science* 68: 935–940.
8. Wright, G. A., et al. 2013. Caffeine in floral nectar enhances a pollinator's memory of reward. *Science* 339: 1202–1204.
9. See Ashihara H., and Suzuki, T. 2004. Distribution and biosynthesis of caffeine in plants. *Frontiers in Biosciences* 9: 1864–1876. The alternative explanation is that the ancestral plant species to all of these plants was able to make caffeine, but this ability

was lost multiple times in different lineages. Because the hypothesized loss events had to occur many more times than the hypothesized evolution of the ability to make caffeine, it is less parsimonious and therefore less likely to have happened. As a rule in science, if we have to choose between two possible explanations without direct evidence for either, we choose by default the simpler (e.g. most parsimonious) explanation.

10. The cacao plant makes a bit of caffeine but mostly theobromine. It may be a case where the third methyltransferases is still evolving and is not yet optimized to methylate theobromine to give caffeine. Or perhaps theobromine is the better defense chemical against the enemies of the cacao plant. Note that when only two methyl groups are added to the three nitrogen atoms present on the xanthine group that are typically methylated, a total of three compounds with the same chemical formula but different structures can be created. These compounds are called "isomers" and all three have been found in plants, but only theobromine is important as a flavor compound.
11. A precursor of the modern sweetened, caffeinated soda drinks.
12. For a general historical description of these wars, see Hanes, W. T. III and Sanello, F. 2002. *The Opium Wars: The Addiction of One Empire and the Corruption of Another.* Sourcebook Inc.
13. Facchini, P. J., and De Luca, V. 2008. Opium poppy and Madagascar periwinkle: model non-model systems to investigate alkaloid biosynthesis in plants. *Plant Journal* 54: 763–784.
14. As we will see in Chapter 6 in the section on rubber, other suspended chemicals can give latex a white color.
15. In fact, prolonged usage of opium causes constipation, since it blocks the activity of the gut muscles.
16. It also benefitted the Indian economy and further benefitted the British economy by stimulating trade between India and England, with the wealth generated in India from the opium trade used to buy finished cotton goods from England. India, which had been a manufacturing center of finished cotton goods, now became a consumer of such goods which were being produced in England at much lower costs because of industrialization and the use of slave labor in America (see Chapter 3). The British government also greatly benefitted from taxing the opium trade in India.
17. A large flow of silver came into China via the trade with the Spanish in the Philippines. The Spanish obtained huge amounts of silver from their mines in South America, but the English did not have abundant sources of silver.
18. Despite their declared neutrality, the American forces intervened to help the British and French in the battle.
19. This Lord Elgin, the 8th, was the son of Lord Elgin the 7th, known for removing the Parthenon Marbles from the Acropolis in Athens and shipping them to England, where they are still on display in the British Museum. The British Museum also holds many artifacts looted from the Summer Palace in Peking.
20. Markel, H. 2011. *An Anatomy of Addiction.* Pantheon Books.
21. *Cannabis* plants are dioecious, meaning that some plants have only male flowers and other plants bear only female flowers.
22. Similar to the ill effects of Prohibition of alcohol consumption in the United States, also at the beginning of the 20th century (1920–1933).
23. Coca-Cola drinks contained traces of cocaine until 1929.

6 Wood and Rubber

WOOD

So far we have seen that people use plants as food, spices, and drugs, all internally consumed. And as the discussion on cotton already showed, plants also provide us with materials to improve the physical conditions of our lives. Cotton fibers, made of the glucose polymer cellulose, are the most common natural material we use to make our clothes. But there are two other plant materials that have significantly shaped the physical aspect of our lives. The first of these is wood, which probably had been important to the human lineage even before it gave rise to the extant species *Homo sapiens*, while the second one, rubber, has only recently begun to play an inordinate role in our lives.

Although it is difficult for us to appreciate the significance of wood to human history, living as we do in an era sometimes called the "Plastic Age," wood has always played a central role in human affairs, on par with stone and metals, and it still does today. From the Stone Age to the Bronze Age to the Iron Age and up to the present, most of humanity has lived in homes that were constructed completely or substantially with wood and that were furnished with wood furniture such as tables, chairs, and beds. Until very recently, our agricultural implements were made wholly of wood or a combination of wood and stone or wood and metal. Basic industrial processes such as oil pressing and weaving used wooden machines. For transportation, we used wooden carts and carriages with wooden wheels, and to cross a river we often went over a wooden bridge. On water, we traveled in wooden boats and ships. In fact, even the first self-propelled airplanes, just about a hundred years ago, were constructed mostly of wood. There is a reason why we name certain periods in history as the Stone Age or the Bronze Age but no period has been called the Wood Age – because the entire history of our species has been a Wood Age.

In addition to serving as construction material, wood has had another major role in human life – as the main form of energy, again a role that has not completely disappeared from even the most technologically developed societies. People around the world used, and some still use today, wood to warm up their houses. Heat from burning wood was, and still is, used for cooking. Wood, and later charcoal, which is prepared from wood, was burned to generate the heat to smelt metals (Chapter 8). On sugar plantations, wood was burned to evaporate the water so the sugar would crystallize. When the steam engine was invented, wood (or coal, which, as we will see in Chapter 8, is fossilized wood) was used to generate the steam.

Of course with so many versatile uses for wood, people were sure to find ways to use it to kill each other. And they did. But before we launch into a description of the many nefarious purposes to which wood has been put to use by people, let us first see what wood is.

Botanically speaking, wood is the hard material that forms the vascular tissues of the plant, where water and nutrients move in the two parallel conduit systems called xylem and phloem (Chapter 3). These systems of pipes are found in all parts of the plant – roots, stems, and leaves.[1] Xylem conducts water and dissolved minerals from the ground up, while organic material moves in the phloem in both directions. For example, metabolites made in the leaves typically move to the roots via the phloem to nurture root cells. But roots can also store nutrients and send them back up to the stem in early spring, as happens with deciduous plants, so that new leaves can grow. Transport of nutrients from roots to shoots can also happen in any plant when above-ground parts are cut or damaged.

Xylem and phloem tissues develop from a type of cell called cambium. The cambium cells divide, and then the new daughter cells undergo a process we call differentiation, in which they no longer resemble their cambium parent cell. The leaves of most plants reach mature size within a few days or weeks, and then stop growing. On the other hand, plant stems, including the main trunk which many trees have, continue to grow in girth during the entire lifespan of the plant. In the stem, differentiated cells become xylem cells on the inside of the cambium cell layer (toward the center of the stem), and phloem cells develop on the outside. Xylem cells eventually die and become hollow, so the water simply moves through them, and through pits that connect the cells, by capillary action. Phloem cells continue to live and the solution of nutrients moves from one phloem cell to the next one in a process that requires metabolic activity and the expenditure of metabolic energy.

The cambium cells in the stem continue to divide sideways throughout the year to give rise to xylem and phloem cells, parallel to the existing xylem and phloem vasculature. In the temperate zone, growth is faster in the spring and it slows down in late summer, giving rise to the famous tree rings – the growth of each year generates two rings, a light, wider one during the spring and early summer and a darker, narrower one indicating the slowing of growth toward the end of the growing season. As the cambium cells continue to divide, the stem grows in girth, with the inside being mostly dead xylem cells. Eventually, the inside of the stem consists completely of dead cells, and this "dead zone" continues to expand outward. The dead center is referred to as "heartwood," and it often has a darker color (Figure 6.1). The dark color is the result of the deposition of a resin of various plant chemicals such as tannins and terpenes[2] which serve as wood preservatives, since they are toxic to microorganisms that would otherwise live off the dead plant material and degrade it. It appears that the synthesis of such compounds by the plant cells is the last thing that occurs before these cells die and turn into "deadwood" heartwood. Basically, the tree embalms its own dead cells in the center of the trunk to preserve them.

Once the heartwood is formed by the impregnation of the toxic, preservative resin, it can no longer conduct water. It does continue to have a function – it physically supports the weight of the rest of the plant above, so that the plant can continue to grow upward and compete for sunlight with neighboring plants, and of course continue to sexually reproduce. The area outside the heartwood zone is called the sapwood, and this is the area responsible for conducting water in the main trunk or in old side branches.

Like all plant cells, xylem and phloem cells cover themselves with a cell wall. Whereas some plant cells, such as the trichomes on the cotton seeds (Chapter 3), have cell walls that are composed almost purely of cellulose fibers, these cases are

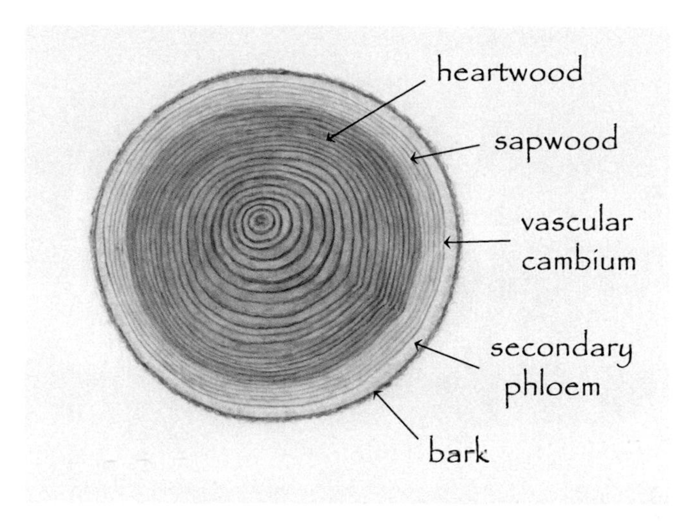

FIGURE 6.1 Cross-section of a tree trunk.

rare. The majority of plant cells have walls that are more complex and contain at least two major ingredients, cellulose and lignin. As we have seen already, cellulose is a long linear polymer of the sugar glucose. Lignin, on the other hand, is an irregular polymer of three main units called H, G, and S, all derived from the amino acid phenylalanine (abbreviated Phe) (Figure 6.2). The plant cells make these units by enzymatically removing the nitrogen atom from Phe, reducing the acid part of the molecule to an alcohol group, and adding more alcohol groups to the phenyl ring (an alcohol group, called by chemists "an hydroxyl" and denoted by the chemical notation –OH, consists of an oxygen atom which is bound by a one covalent bond to an hydrogen atom, and which has the capacity to form another covalent bond). To form lignin, the plant cells link H, G, and S units to each other via their oxygen or carbon atoms to form a three-dimensional lattice (Figure 6.2), meaning that it is not a simple linear polymer, but each unit can be linked to more than one other unit.[3] Moreover, while we still know very little about how plants accomplish this task, it appears that there is no regularity to this lignin polymer, so that any H, G, or S unit can be linked to any other and via different parts of the molecules in a seemingly random fashion. The ratio of the different units can also vary even in different organs of the same plant. However, the S unit appears to be lacking in the lignin of gymnosperm species, while it is usually the dominant unit in the lignin of angiosperm plants.

The lignin polymer forms a hard material that is quite strong by itself. But what gives wood its exceptional strength to resist compression, tension, and shear forces is the combination of lignin with cellulose. Plant enzymes chemically cross-link the two polymers, again using the oxygen atoms in both types of polymers as the "bridges" (Figure 6.2). The composite polymer that results from this cross-linking, which in heartwood is also impregnated with resin compounds, is what we call wood, since in older trunks and stem very little other organic material is left. This material possesses extreme strength and flexibility, capable of supporting tons and tons of weight (think of a huge sequoia tree) as well as side-wind force. It is also relatively

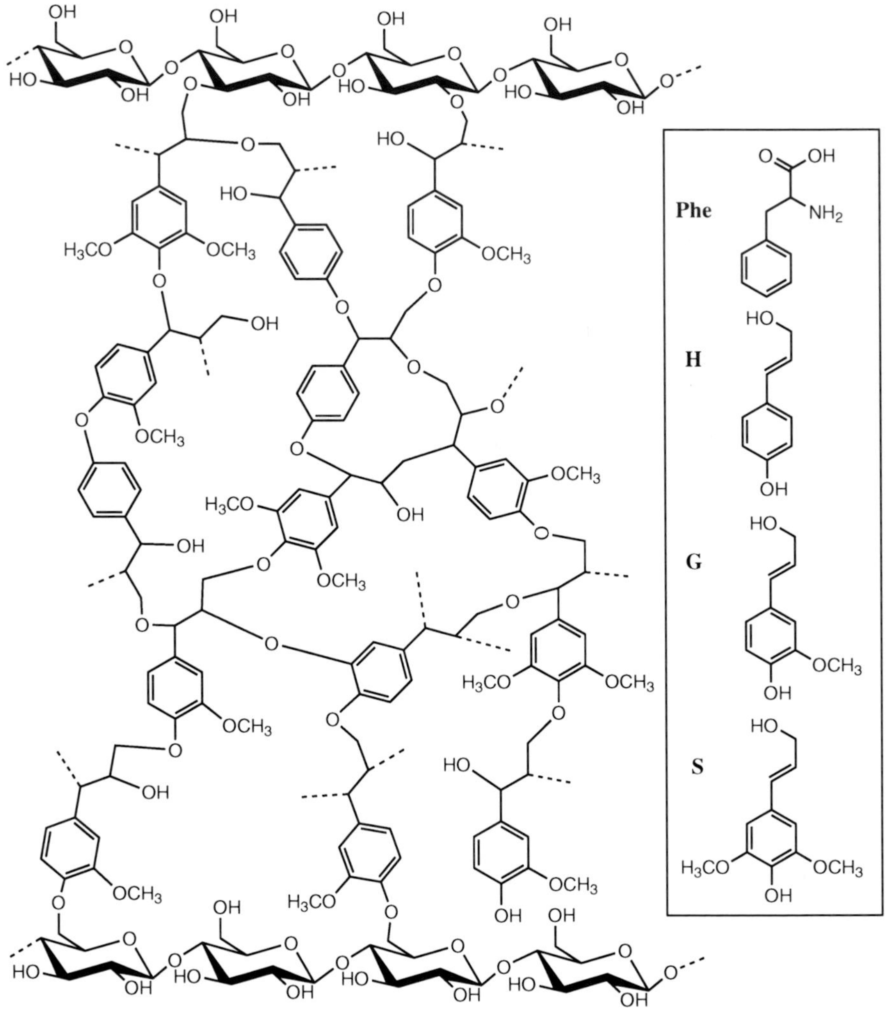

FIGURE 6.2 A schematic showing the structure of the irregular polymer lignin in the cell wall and its linkage to the cellulose linear polymers. Inset, the three monomers of lignin, called H, G, and S, all derived from the amino acid phenylalanine (Phe). The lattice is shown here in two dimensions, but it should be borne in mind that it is a three-dimensional lattice, and dashed lined indicate linkage to additional molecules in all directions.

light weight (compared to steel, for example) and, depending on the amounts of natural plant toxins present in it, it can be highly resistant to biological degradation. Only some fungi and bacteria are able to degrade lignin and cellulose, and compounds such as tannins and terpenes found in wood act to suppress their growth. Dense heartwood often survives for hundreds or even thousands of years without breaking down and losing its strength.

Wood was one of the first materials that *Homo sapiens* used for building tools, and it is likely that even the ancestors of our species used wood, since some of our extant relatives still do. For example, chimpanzees have been observed to use plant sticks to pull termites out of their mound. There is archeological evidence, mostly from stone paintings and carvings, that already in the late Paleolithic era (which ended about 10,000 years ago), humans made wooden clubs and fashioned spears from long, slender shoots that they sharpened at the end. Such weapons were possibly used first for hunting. Bow and arrow technology, based on the tensile properties of wood imparted by the cellulose-lignin mixed polymer, is also extremely ancient. An 11,000-year-old wooden bow made from elm was found in a swamp in Denmark, and arrows made from European hazel tree and wayfaring tree, dating to about the same period, were found elsewhere in Europe. With the exception of the Australian aborigines who were separated from the rest of humanity since their arrival in Australia 50,000 years ago, all other indigenous people on earth are known to have used bows and arrows, attesting to the likely single and very ancient origin of this weapon system.[4]

WOOD IN LAND WARFARE

While bows and arrows were likely first used for hunting, they soon became a major weapon system employed in warfare. Recorded history of practically any culture around the world provides evidence for their use in battle. Various versions of bows were soon constructed, including the crossbow and composite bow. The crossbow, invented in China by 500 BC, is a bow mounted on a wooden board (called a stock). The pulled string of the bow is held by a trigger mounted on the stock, so the archer can hold the stock and take aim, then release the string by pulling on the trigger. The arrow used by the crossbow, called the bolt, could be tipped with a metal head. Early crossbows were made from a single piece of wood, such as yew or ash. Later, composite crossbow were developed, with wood in the middle, horn material on the inside, and sinew on the outside.

The major weapon used by the Mongols under Genghis Khan and his heirs, who established the largest empire known in history (by territory) in the 13th century, was a composite bow, about 80–100 centimeters in length and made with bamboo wood in the middle flanked by horn on one side and sinew on the other. The arrows were made from birch wood. The Mongol warrior was mounted on a horse and thus highly mobile, and was very proficient in firing his arrows in motion.

Another military tactic that used bows and arrows relied on the longbow. The English longbow was made from a single piece of wood, typically yew, but took advantage of the difference between sapwood and heartwood. The wood was cut in such a way that along its width, about one-third was sapwood, which is good in tension, and two-thirds hardwood, which is better at resisting compression. The sapwood part was on the outside of the bow (away from the archer). In essence, the English longbow was made of natural composite wood. The English longbow was about 1.8 meters long, and the metal–tipped arrows, made from poplar, ash, beech or hazel wood, could be shot as far as 300 meters. The longbow was too big to be used by a mounted archer.

Longbows were a main weapon in Europe during the medieval period, and because of the demand for wood to make them, yew trees became very scarce there.

The longbow played a crucial role in the success of the English forces in several battles during the Hundred Years' War between England and France (1337–1453), a war that was fought over the claim of the English kings to the French throne (due to some complicated family history). The most famous was the Battle of Agincourt in 1415, where the small army of Henry V, with 9,000 English soldiers, most of them longbow archers, defeated a larger force of French soldiers. The French employed few archers but had several thousand knights in armor and a large cavalry force. Since the English longbow archers needed to be stationary on the ground to operate their bows and thus in a vulnerable position, they successfully protected themselves from the cavalry by driving wooden stakes, sharpened on the exposed ends, into the ground around their position. These spikes, and the arrows of the archers, injured or killed many of the horses of the French knights, and in addition the arrows were also able to penetrate weak spots in the knights' armor, particularly in the parts covering the arms and legs. When the French knights, now unhorsed but still wearing heavy armor, finally made direct contact with the English forces after having to walk in a muddy field, they were exhausted and easily beaten.

Agincourt showed that a long-distance weapon made from wood (with the exception of the arrow head, which was made of iron) can defeat heavily armored soldiers. But although bows and arrows continued to be used in major warfare around the world until the late 19th century (see Chapter 5 on their use, although to little effect, by Chinese soldiers during the Opium Wars), Agincourt was the last major victory for the longbow. By that time, simple guns and gunpowder had already been introduced, and they eventually made bows and arrows obsolete.

Wood was also used extensively in defense. Wooden palisades, created by driving logs or stakes vertically into the ground next to each other, have been employed by virtually all human societies, particularly when stone was not available at all and often as a supplement to stone or dirt walls. For example, Amerindian villages in the eastern part of North America were often protected by wooden palisades.

Until the use of cannons became commonplace, most siege machines were constructed at least in partly from wood. While all dominant human cultures became so by waging aggressive warfare, by invading neighboring countries, and by laying siege to their cities, the ancient Romans in particular made siege war into a science and developed a huge list of siege machines. Their war machines were made mostly from wood, and included such implements as the ballista, the onager (both for throwing projectiles), the siege tower, siege ladders, and the battering ram. The latter was made from a very large wooden log, suspended from a tall frame by ropes made of plant material, and covered at one end with a metal cap. The log was swung back and forth by soldiers, and its end, sporting an iron cap sometimes fashioned as ram head, was made to hit the wall with as much force as possible.

A detailed description of the famous Roman siege of Jerusalem in 70 AD was provided by the historian Josephus Flavius, an eyewitness to the event.[5] Titus, the son of Emperor Vespasian (and eventually an emperor himself), commanded the Roman troops putting down the Jewish rebellion. Jerusalem, protected by three concentric stone walls, was one of the last holdouts against the Roman army. The Romans employed siege towers, battering rams, and artillery machines, breaching the first wall and then the second wall and eventually pushing the defenders back into the temple area. Up to this point the defenders were courageously fighting the Romans by taking advantage of the biggest drawback of wood – its flammability. The parts of the siege devices facing the defenders were protected from flammable projectiles with iron plates, but the Jews occasionally sortied out of the walls and set the towers and the battering rams on fire anyway, occasionally causing serious damage to the equipment. The Romans invariably repaired the machines or replaced them, continuing their slow advance. Once the Romans reached the inner temple, the fighting conditions changed. This area was high up on a hill and it would have taken the Romans a while to bring the siege towers and the battering rams up there. Instead, the Romans attempted to scale the wall with wooden ladders, but some Roman soldiers became trapped in a small area inside the temple compound encompassed by wooden walls, which the defenders set on fire, causing the Roman soldiers to be burned alive. Titus then decided to set fire to the wooden gates in the wall guarding the temple and eventually – apparently against the orders of Titus – some Roman soldiers also set fire to the temple structure itself, also mostly made of wood. The huge conflagration allowed the Romans access to the temple area, and they proceeded to massacre the remaining Jewish defenders, while the temple was completely destroyed in the fire.

WOOD IN SEA WARFARE

Wood as a weapon was just as important at sea. The ability of most wood to float on water no doubt led ancient people to realize that they could use simple wood rafts to transport themselves and their cargo across bodies of waters. While no archeological evidence remains, the Australian aborigines must have used some form of boats or rafts to reach Australia about 50,000 years ago. There is firm hieroglyphic evidence for the development of hulled wooden vessels almost 5,000 years ago in Egypt. A complete ship made of cedar tree (*Cedrus libani*) wood (Figure 6.3) and measuring 44 meters in length was found buried by the Great Pyramid of Giza, and was dated to around 2500 BC.

Ship building spread around the world early. As a form of conveyance, wooden ships were used to establish new colonies and for commerce between different

FIGURE 6.3 Trees used for ship building: From left, Cedar of Lebanon (*Cedrus libani*), English oak (*Quercus robur*), and pine (*Pinus strobus*).

countries and human societies. But they were also used to convey soldiers to military targets and brigands to attack civilian populations. With the presence of merchant ships in open sea, a new form of violent crime developed – piracy. Piracy, whose ancient origins are attested to in many texts such as Homer's *Iliad* and *Odyssey*, is strictly defined as robbery and violent acts against non-combatants at sea, although it has often been used to describe such acts committed on lands by criminals arriving by boat. Piracy has thrived at one time or another practically everywhere in the world, particularly in places and times that lack a strong central government. For example, piracy was rife in the China Sea during the weakening of the Ming dynasty in China in the 16th and 17th centuries, and in the Straits of Malacca after the Portuguese conquered Malacca but did not have the naval force to patrol the area. In some cases, governments actually commissioned some private ship captains to commit piracy against ships of other countries, as the Englishman Francis Drake was directed to do by his queen, particularly against Spanish ships anywhere he encountered them.[6]

But throughout history, nowhere had piracy been more prevalent, or better documented, than in the Mediterranean Sea. The Greeks actually considered piracy a noble military enterprise, on par with raiding and sacking cities (other than one's own) by land forces. As a result, Greek ports were not only havens for pirates but they also had big slave markets (selling captured passengers as slaves or ransoming them were the two major sources of revenues for pirates). The Romans began a successful effort to fight piracy in the 1st century BC, and thereby ensured safe commerce on the Mediterranean for several hundred years. With the decline of the Roman Empire, piracy returned in force to different parts of the Mediterranean coast, inhibiting commerce and forcing many countries either to battle pirates or pay tribute to protect their merchant ships. The young United States of America was one of them. While the United States initially paid tribute to the pirates of the Barbary Coast, situated between present-day Libya and Morocco, President Jefferson refused to pay and sent a squadron of US Navy ships to bombard coastal ports and destroy pirates' ships. These efforts were only partially successful and a treaty signed in 1805 was soon violated by the pirates. President Madison sent more navy ships in 1815, and again a treaty was reached with Algerian pirates but it too was not honored. It took the involvement of the English and Dutch navies the next year to stop most piracy, although it did not completely end until France invaded Algeria and made it into a colony in 1830.

With the fear of having ships attacked at sea, and of enemies arriving from the sea to attack cities, ships that could respond to such aggression were developed. Eventually, warships engaged each other in sea battles, which were some of the most spectacular and deadly military engagements in history, with casualties that far exceeded those of land battles, both in total numbers and as a percentage of combatants involved.[7] The Greeks in the 7th–4th centuries BC developed highly sophisticated warships. The biggest battleship of the time was the trireme (probably of Phoenician origin), about 35 meters in length and six meters wide. The name refers to the three horizontal rows of oars, one on top of the other, and on both sides of the vessel, for a total of approximately 170 rowers. The triremes were made from relatively light wood such as fir, pine or cedar, so that the ship could be easily pulled out of the water to dry out or for repairs. However, oak wood was sometimes used in the hull construction, because such wood was denser and more able to withstand damage when the ship was beached. To prevent water seeping through the seams where the boards were connected, pitch, the resin made of terpene compounds exuded by conifer trees such as pine and cedar, was used to caulk the hull. Taut ropes made from papyrus and flax extended from fore to aft, giving the ship additional rigidity. The ship could be fitted with fir or pine masts but was mostly oar-propelled.

The Greeks developed two tactics for naval warfare that remained practically unchanged for more than 2000 years, until the addition of heavy guns to ships starting in the 15th century AD. The first was ramming the ship into the enemy ship, aiming to hit it in its middle and crashing into it at a right angle. For this tactic to work, the attacking ship had to be fast and maneuverable. It could therefore not rely on wind power, but instead speed and control were achieved with oars. The bow was equipped with a built-in ram that could puncture the hull of the enemy ship at or below the water level. Ramming the enemy ship often sufficed to cause it to break up and sink. Colliding sideways into an enemy ship was also often attempted with the purpose of breaking its oars (after withdrawing one's own oars first). Once physical contact between the two ships was made, the second tactic of Greek naval warfare was employed – boarding the enemy ship with armored soldiers (called "hoplites") and engaging in hand-to-hand combat.

The most famous and consequential naval battle in Greek history was the Battle of Salamis in September of 480 BC, during the second Persian invasion of Greece. Earlier in the year, the Persian king Xerxes I and his huge army crossed the Bosphorus on two bridges, each about 1,500 meters long and made from hundreds of ships lashed together with papyrus and flax ropes. His army then proceeded to conquer Greece from the north toward the south. Soon, the Persians arrived in Athens. According to Herodotus, the Athenian leader Themistocles had anticipated the Persian invasion and believed that the best way to fight them would be at sea, where the Athenians were strongest. He then convinced the Athenians that an earlier oracle saying that Athens will be protected by "wooden walls" should be interpreted to mean that Athens needs to build a lot of ships, as they were the "wooden walls."

Athens indeed prepared a navy of 200 triremes, and when the Persians approached Athens, the Athenians evacuated the civilian population to the nearby island of Salamis, separated by a narrow strait from the mainland part of the Athenian territory where Athens was, and the Athenian men – free citizens, not slaves – withdrew to their triremes. They kept their triremes on the beach of Salamis, and there the Athenian ships were joined with the ships of their Greek allies, with the complete fleet having about 370 triremes. Themistocles, using a ruse, was then able to lure the navy of the Persians and their allies, estimated to be between 600–1200 ships, into the straits between Salamis and the mainland. In the battle that ensued, the Greeks won a decisive victory, destroying or capturing hundreds of Persian ships and losing only 40 of their own ships. This battle was the beginning of the end of the Persian invasion. Afterward, Xerxes withdrew most of his forces from Greece, and the remaining Persian forces were subsequently decisively defeated in additional land and sea battles the following year.

Naval battles along the line exemplified by Salamis were fought in the Mediterranean Sea for the next 2,000 years. The last major battle in which oar-powered warships fought each other was the Battle of Lepanto, at the mouth of the Gulf of Corinth on the western side of Greece, in 1571 AD. The battle was between the forces of the Christian coalition of European states against the navy of the Ottoman Empire, each side numbering a little over 200 galleys (a later version of the trireme, with three banks of oars and some sails) and a few other types of ships. By this time, heavy guns (cannons) had already been added to some European warships, and some European soldiers were equipped with rudimentary personal guns. Due mostly to the advantage of their guns – the Turks still relied mostly on longbow and hand-to-hand combat – the Christian forces won a decisive victory that basically put an end to the Ottoman Empire's naval dominance in the Mediterranean.[8]

Guns mounted on ships, with their metal or stone projectiles, could hit and sink the wooden enemy ships from a distance, without the other side ever having a chance to fight back. This obvious advantage, made clear at Lepanto, led to the addition of more and more guns on board, with some ships carrying more than a hundred guns. To carry that amount of weight, ships had to be made bigger and sturdier, and they could no longer be rowed by people. Until the introduction of steam engines, warships had to be powered by wind and therefore equipped with big masts. To make them stronger, the ships were now being constructed with oak wood, mostly the English oak, *Quercus robur*, a tree species that grew in abundance over most of Europe.[9] The ships were fitted with fir or pine masts, since these coniferous trees grew tall and had straight trunks. Five hundred years ago Europe was much less populated than today and still mostly forested, but the demand for wood for ship building was soon to become a major cause of deforestation. By the 18th century, construction of a ship of the line (see below) required between 3,000–4000 oak trees. North American oaks, such as the white oak (*Q. alba*), were soon tapped for ship building and even the American red oak (*Q. rubra*), considered inferior in quality, was used as a last resort.

Ship building was also very expensive, and one of the main causes for the Civil Wars in England that began in 1642 was due to the resentment of the "Ship Tax" imposed by King Charles I in 1634.[10] Traditionally, such a tax could be imposed by the king without parliamentary approval, but only during war and then it applied only to coastal residents. But while England was at peace in 1634, the king wanted to establish a permanent fleet of warships, and thus made the tax permanent and applied it to all inhabitants of England, an action that generated much resistance. The need to obtain wood for ship building got the English kings in trouble with their subjects living in North America as well. When big pine trees for masts became scarce in Europe, the English kings decreed that all eastern white pine trees (*Pinus strobus*)[11] growing in their American colonies within ten miles of a navigable waterway, and having a trunk wider than 24 inches, belonged to them and could not be used by the colonists. By 1772, the English Parliament extended the ban to trees with a diameter of at least 12 inches. The enforcement of this ban by the English authorities led to military skirmishes, called The White Pine War, with settlers on whose properties such trees grew, and, by adding to the resentment of the settlers, this ban contributed to the settlers' final rebellion against English rule a few years later.

The 19th century was the last century in which wooden warships played a major role in naval battles. By the end of the century, they were replaced by ships with metal hulls and steam engines. The mechanical propulsion system – strong, reliable, constant, and not dependent on the fickle wind – was an obvious advantage. The metal hulls not only gave structural strength to the ships, but also protected them from the two main weaknesses presented by wood. Wooden ships were susceptible to fire, particularly once gunpowder was stored on board. In addition, the wooden hull was often infected with shipworms, which, despite their names, are actually mollusks belonging to several genera that burrow into the wood while consuming it,[12] thereby weakening the wood and often creating outright breaches that allow water in. With heavy warships staying in the water for very long periods of time, shipworm infection was a problem that was never satisfactorily solved, with many ships springing major leaks and often sinking as a result. The problem was exacerbated with ships that spent more time in warmer waters where the mollusks grow faster, a situation often encountered by the navies of the European colonial powers.

But at the beginning of the 19th century, wooden ships still controlled the high seas. The century began with the famous battle of Trafalgar, which both showcased the by then well-established standard tactics of naval warfare as well as a radical departure from such tactics. It was customary for two opposing navies to form two long parallel lines of ships (hence the term "ship of the line"), with the ships of each line facing the opponent's ships sideways, and then to shoot at each other with their guns (hence, a "broadside") until the opposing ships were sunk or until one side decided to leave. In 1805, off the Atlantic coast of Trafalgar at the south-west tip of Spain between Cadiz and Gibraltar, the 33 ships commanded by English admiral Nelson encountered a combined French–Spanish force of 41 ships. Once the two armadas saw each other, the combined French and the Spanish forces indeed formed a long line. Nelson, however, departed from this standard tactic and instead divided his force into two columns that sailed perpendicularly to the enemy's column, breaking it into three separate groups and then encircling some of them and fighting in close contact. The plan worked well, cutting off the lead section of the enemy ships, which could not turn around because of the wind conditions and thus did not take part in the ensuing battle. However, during the initial, perpendicular approach toward the enemy ships, the English ships could not employ their guns, which were mounted on the sides of the ships, and Nelson himself was hit by a musket bullet. He died a few hours later, having been informed of victory. The Battle of Trafalgar, occurring during the Revolutionary Wars, removed any hope that Napoleon may have nursed of invading England. The war went on, however, for ten more years.

A side story of the Battle of Trafalgar illustrates another wood technology, the barrel. The bodies of sailors dying at sea were typically disposed of at sea, but the bodies of more senior commanders, such as Admiral Nelson, were transported for proper burial on land. To prevent decomposition, Nelson's body was placed in an oak barrel – wooden barrels being the main technology for storing and transporting liquids at the time – filled with brandy in which camphor and myrrh were dissolved. The plant-derived ethanol (the brandy) prevented bacterial growth, and the plant-derived terpene defense compounds camphor and myrrh exerted their own antimicrobial toxic effects. The barrel was placed on Nelson's ship Victory while it was towed to Gibraltar for repairs. From Gibraltar, his body was transported to England in a lead-lined coffin filled with distilled spirits of wine (again, ethanol).

RUBBER'S GOOD QUALITIES

Today, a huge number of man-made objects are constructed wholly or partially from the plant material rubber.[13] Waterproof boots and raincoats, elastic bands in many clothing items, hoses and balloons, sports balls, medical equipment, surgical gloves, condoms, various components of electric equipment, various components of many machines – the list goes on and on. It has been estimated that rubber is found in over 40,000 modern products. Above all, today 80% of all rubber obtained from plants goes to making tires and inner tubes for the wheels of bicycles, cars, trucks, airplanes, and many other transportation devices (with roughly 40–50% of tires made of natural rubber).

The widespread use of rubber[14] is due to its remarkable combination of physical and chemical properties, unmatched by any other natural or synthetic material. Rubber is elastic, and it will return to its original shape when either pulled or squeezed by the very strong forces that could break or crush any other non-rigid natural material. It is watertight and airtight. It is a good electric insulator. Volcanized rubber is strongly resistant to abrasion and caustic chemical agents such as acids. And it originally comes from plants as very small particles suspended in a liquid, so it can easily be molded to form any shape.

Rubber is a linear polymer of the five-carbon compound isoprene, which, as we already saw in Chapter 4, is the basic unit from which terpenes are made. When only two to four units of isoprene are joined together by the plant's enzymes, the terpenes produced have between 10–20 carbons, small enough to be volatile. Thus, many such small plant terpenes are used by people as spices. All plants also make polymers of isoprene that have up to 20–30 or so isoprene units. These polymers are generally called polyisoprenes. They are hydrophobic molecules not soluble in water and are found in the membranes of plant cells, serving various cellular functions.

Some plants – more than 2,500 species have been found so far – have also evolved enzymes that can link hundreds and even thousands of isoprene units to form a very long polymer. The double bond in the basic isoprene unit makes it a rigid, flat molecule, with all the carbons being on the same plane. Rubber is made when the isoprene units are joined together in such a way so that the backbone of the complete polymer (the chain of carbon molecules linked to each other) always connects to the two outside carbons on the same side relative to the double-bond position, an arrangement called a *cis* configuration[15] (Figure 6.4). When the units are joined together in the alternative way, the *trans* configuration, the polyisoprene that is produced is not rubber but the material called gutta-percha, or balata. Rubber and gutta-percha are typically made in specialized plant cells called laticifers[16] (discussed also in Chapter 5). The rubber or gutta-percha made in these cells are not water-soluble and are too big to remain in the cell membranes. Instead, they form tiny particles that are suspended in the cells' watery cytoplasm, hence this cytoplasm is called latex. Note that, unlike common use, the term latex in botany is not synonymous with rubber but connotes any plant sap that has colloidal particles in it and therefore, in the absence of any pigment, appears white and not clear.

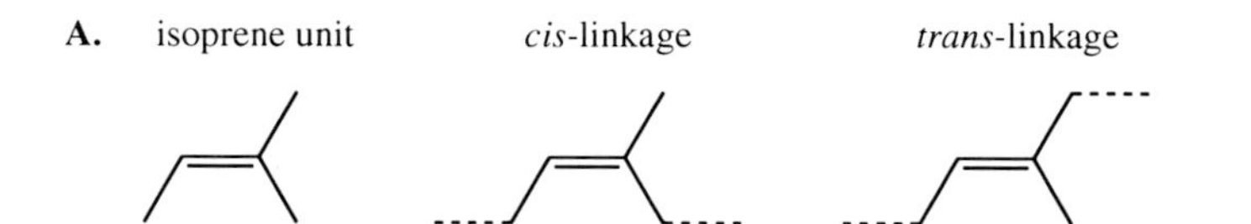

B. rubber (*cis*-polyisoprene)

C. vulcanized rubber (*cis*-polyisoprene with sulfur cross-linking)

D. gutta-percha (*trans*-polyisoprene)

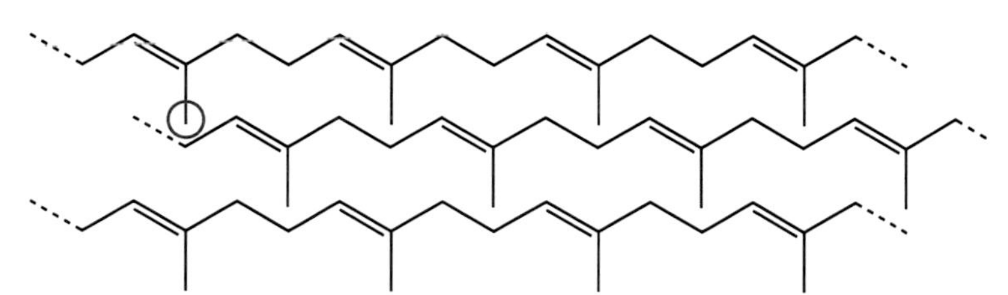

FIGURE 6.4 (a) The structure of the basic subunit of rubber and gutta-percha, the isoprene, and a demonstration of what constitute *cis* and *trans* linkages between two isoprene units. (b) The structures of natural rubber, the isoprene polymers in which all (or almost all) isoprene subunits are linked in *cis*. (c) The structures of vulcanized rubber, showing the cross-linking between linear polymers with sulfur atoms. (d) The structures of gutta-percha (also called balata), the isoprene polymers in which all isoprene subunits are linked in *trans*. The red circles indicate examples of methyl groups.

The rubber or gutta-percha particles in the latex can constitute anywhere from traces up to 50% of the total volume, as in the para rubber tree, *Hevea brasiliensis*. In addition, the latex of rubber or gutta-percha producing plants typically contains several other defense reagents, such as proteins that interfere with the digestion of plant material that herbivores ingest, and also enzymes that can digest the chemical chitin, which gives strength to the skin of insect and the cell wall of fungi.[17] Thus, when the stems or leaves of plants containing rubber or gutta-percha are physically damaged by a herbivore, the latex comes out and can damage the offending animal on surface contact or when ingested by the animal. The rubber and gutta-percha materials themselves are initially soft, but they are sticky and so they start immobilizing the animal right away as well as suffocating it, the more so as the water in the latex evaporates. Finally, rubber and gutta-percha are also excellent at sealing the wounds to the plant body caused by the herbivores and preventing infection by bacteria and fungi. Thus, the natural function of rubber and gutta-percha appears to be similar in principle to the resin of conifer trees, made mostly of sticky diterpene acids (four units of isoprene), which oozes out of a wounded stem, seals the wound, and engulfs and suffocates the attacking insects (fossilized terpene resin is called amber).

Rubber and gutta-percha have somewhat different physical and chemical properties. Because of the *trans* configuration of the long isoprene polymer in gutta-percha, the individual chains assume a stretched-out position and they can come into close contact with each other, thus the material as a whole is more dense and less elastic. While fresh gutta-percha is harder than fresh rubber at ambient temperature, it can be warmed up and remolded. However, after a long exposure to air (but not water), the polymer strands undergo an oxidation process that chemically links them to each other, and the material as a whole loses its elasticity and becomes irreversibly hard and brittle. In rubber, because of the *cis* links between the isoprene units, each polymer strand assumes a configuration that is more convoluted, or coiled, with the result that the separate strands have less contact with each other. Rubber therefore has the ability to stretch more that gutta-percha. A crucial aspect of the linear polymers of both rubber and gutta-percha is the *methyl* side groups (each consisting of one carbon with its three hydrogens) that stick out regularly off the backbone of these polymer strands (Figure 6.4). These methyl groups cause the strands to get entangled together as the material is stretched or compressed, thus preventing the strands from sliding past each other. This entanglement gives the materials the strength both to avoid being torn apart and to resist compression. Once the pulling or compressing forces applied to the material – both of which distort the bond angles between atoms in the polymer molecules – are removed, the molecules in these materials resume the correct bond angles because this is the most thermodynamically stable condition, and therefore the material returns to its original shape.

Only very few plants have been found to make gutta-percha (Figure 6.5), and while these trees are found in tropical areas of both the Americas and the Far East (the material is called gutta-percha in the Far East, balata in the Americas), they all belong to the same family of flowering plants, the Sapotaceae. Because of the irreversible brittleness that gutta-percha acquires over time, today this material is used very little. In the past, the natives in Borneo made simple tools from it, and in the middle of the 19th century it was heavily used for insulating undersea telegraph cables, since the material keeps its elasticity under water.[18] It was also used to make a type of golf ball, the "guttie," that was a great improvement over the "featherie," a ball with leather skin filled with feathers, for tooth fillings, and in other sundry artifacts. However, unlike for rubber, synthetic plastic materials with superior quality have been developed to replace gutta-percha in most of its original uses, so today there is little demand for this material.

Rubber-producing plant species are much more common in nature than gutta-percha producing ones. However, the amount of rubber present in the latex and the average length of the polymer strand – the longer it is, the stronger the rubber is – vary greatly among plants.[19] Europeans first encountered rubber in the 15th century when they invaded Central America. The elastic balls used in the ballgames of the Maya and Aztec civilizations were made from rubber obtained from the Panama rubber tree, *Castilla elastica* (Figure 6.5). This tree grows all over Central America, as well as in the northwestern reaches of the Amazon basin. Besides balls, the locals also made waterproof garments and shoes from this rubber. The manufacturing process involved mixing the rubber with the sap of the morning glory plant, *Ipomoena alba*, molding the shape of the item, and letting it set.

FIGURE 6.5 Top from left to right: the rubber-producing plants *Hevea Brasiliensis* and *Castilla elastica*, both from South America, and the African *Landolphia owariensis*. Bottom from left: the gutta-percha (balata) producing plants *Palaquium gutta* from East Asia and *Manilkara bidentata* from the Americas.

Throughout most of the Amazon basin, another tree grows that produces more and stronger rubber than *Castilla elastica* does. This is the Para rubber tree, *Hevea brasiliensis* (Figure 6.5). The average number of isoprene units in the polyisoprene strands that this tree makes is 20,000. There are only two other plants known to have better quality rubber – guayule and Russian dandelion. However, neither of these two species produces nearly as much rubber as *Hevea brasiliensis* does. The latex of the Para tree, which can grow up to 30 meters in height, is up to 50% rubber, and the latex itself comprises 2% of the total weight of the tree. The rubber from the Para tree is also easily accessible – the latex can be harvested by a non-destructive "tapping" procedure in which the tree trunk is superficially cut, and the resulting exudation of latex is collected into a container.

In 1734, the French scientist Charles de la Condamine paddled down the Amazon River from Peru to its mouth on the Atlantic, and during the trip he observed that the locals used rubber from the Para tree to make various items such as shoes and bottles. He took some Para latex with him back to France and used it, among other things, to waterproof his coat. Other French people working in the Amazon experimented with the material, and continued to do so when they came back to France, making balls, erasers, and watertight bags. In the 1790s, the cloth used to construct hot-air balloons, a French invention, was made airtight by impregnating it with rubber. At about the same time, the use of rubber spread to the rest of Europe, with English manufacturers making rubber hoses and machine belts as well as rubberized, waterproof garments. In 1823, for example, the Scottish chemist Charles Macintosh introduced his namesake waterproof raincoat, which was made by treating rubber with heavy hydrocarbons ("naphtha") generated from coal, and sandwiching the resulting material between two sheets of cotton.

The use of natural rubber to make artifacts presents an opposite problem to that found with gutta-percha. Unlike the latter, which invariably becomes hard and brittle over time, rubber usually remains sticky (and smelly). But it melts at warmer temperatures, and in cold temperatures and upon ageing it does become brittle like gutta-percha. Macintosh's initial waterproof clothing, although commercially successful, still suffered from these problems. The solution was discovered in the late 1830s to early 1840s by two people, American Charles Goodyear and Englishman Thomas Hancock (the latter admitted to being influenced by the work of Goodyear). The process they came up with, which Goodyear dubbed "vulcanization," was to use sulfur to cross-link the polyisoprene strands in rubber, a process that gives the rubber additional strength, analogous to what happens when lignin cross-links cellulose strands in wood (see earlier in the chapter). By varying the number of sulfur "bridges" linking the polyisoprene strands and the number of sulfur molecules in each sulfur bridge (the sulfur atoms can link to each other too, forming their own chain), as well as adding other ingredients, rubber with different levels of strength, flexibility, and durability could be obtained. Once vulcanization solved rubber's problem with strength and durability, rubber became a common material in construction of various consumer goods and components of machinery. With the invention of the pneumatic tire, first for bicycles and, toward the end of the 19th century, for cars, the demand for rubber exploded.

RUBBER'S CONTRIBUTION TO HUMAN MISERY

Until the late 19th century, practically all the rubber used by industry came from wild plants. The majority of it came from the Amazon, where *Hevea brasiliensis* and *Castilla elastica* grew[20] (Figure 6.6). These trees were found in the jungle among other trees, irregularly spaced apart, often as much as several hundred meters away from each other. Tapping and collecting of rubber were often done by individuals working as subcontractors to companies that controlled swathes of the jungle. The work was hard and dangerous, both physically and due to tropical diseases. Many of the tappers, called *seringueros*, died within a few years of starting work, and many were in debt to the company that hired them, paid their travel to the jungle, and sold them work equipment and food supplies at exorbitant prices. But stories about *seringueros* making it rich kept workers coming. Certainly the Amazon rubber boom in the second half of the 19th century, which extended until 1910, made quite a few people immensely rich, but usually not the tappers.

Another major source of rubber came from a peculiar African colony personally owned by King Leopold II of Belgium, called The Congo Free State. Until late in the 19th century, European countries had only colonized Africa on its edges, and had not yet penetrated into the tropical center of the continent. The major reason holding them back was the ubiquitous presence of tropical diseases, particularly sleeping sickness spread by the tsetse fly, for which Europeans had no immunity (and which also devastated the locals). King Leopold II of Belgium, seeing how profits from the colonies of his Dutch neighbors were making the Dutch State prosperous, strongly felt that Belgium should have its own colony. He set his heart on the Congo, which had two major exploitable resources. The first was ivory from elephant tusks. The second was a type of red-tinted rubber that could be extracted from various vines of the genus *Landolphia* growing in the jungle.

Leopold II was quite cunning. Belgium was not a major military power, and besides, the Belgian government was not actually interested in acquiring an African colony. So to achieve his goal, he spent several years in preparation, during which time he set up a bogus humanitarian committee tasked with looking at ways to improve the lives of the people of the Congo.[21] Then, at the 1885 Berlin Conference,

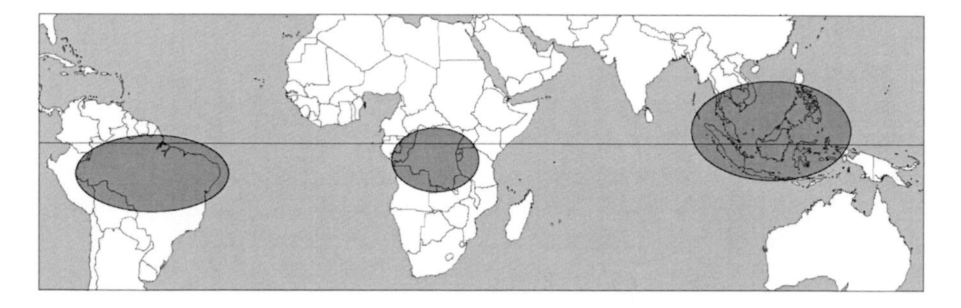

FIGURE 6.6 The tropical regions of the world where wild trees and vines producing abundant rubber and gutta-percha originally grew.

convened by the German Chancellor Otto Bismarck specifically to carve up large parts of Africa among the European powers, he convinced the other participants – representative of European countries as well as the United States of America – to grant him personal control of the area of the Congo. He called his new fiefdom the Congo Free State.

Once given the deed to the land, Leopold II proceeded to turn his "free" state into a forced labor camp. Since the state belonged to the king and not to the State of Belgium, the laws of Belgium did not apply, and the king had complete control to do as he pleased. To make sure the locals obeyed him, he set up a private army called Force Publique, composed of European officers and African soldiers recruited from outside the Congo. The locals were no longer allowed to sell ivory and rubber (or anything else) to private companies, and instead they were forced to sell them to the state, which set the prices. When the natives did not bring enough ivory and rubber to the officials, they were assigned quotas. Failure to deliver the assigned quota resulted in beatings, hostage taking, beheadings, rapes, and various other forms of torture and terror at the hands of the Force Publique. In particular, cutting off hands as a punishment for lack of productivity became widespread. Local rebellions against this harsh rule were similarly brutally suppressed.

This system of extreme exploitation and cruelty, which eventually developed into a concessionary system run by companies controlled by friends of Leopold II or directly by him, continued until 1908. By 1900, 85% of the value of the products extracted from the Congo was in rubber, and it made Leopold and his friends very rich. The effect on the local population, on the other hand, was horrendous. Reliable estimates indicate that perhaps as many as 10 million Congolese died during the reign of Leopold II. Many Congolese escaped to neighboring regions, and the country became depopulated, with the population decreasing by as much as 50%.

The atrocities going on in the Congo were not exactly a secret to the rest of the world. Some European visitors who witnessed them wrote about them when they returned to Europe. Joseph Conrad, a Polish ship captain who piloted a boat on the Congo River in 1890, later settled in England and published his novella *Heart of Darkness* in 1899 describing, in fictionalized form, some of the horrible things he had seen. In 1903, the British government, yielding to English (and American) public pressure, appointed its consul in Congo, Roger Casement, to carry out an investigation. Casement, who was an old Congo hand, conducted a thorough on-site examination and delivered the Casement Report in 1904, which confirmed the worst allegations. Based on this report, the English government demanded to reopen the 1885 Berlin Conference agreement giving Leopold II the right to govern the Congo. The Belgian government resisted, but prevailed on Leopold II to set up his own "independent" commission, whose members, to the amazement of everyone, including Leopold II, in 1905 essentially confirmed the basic facts of the Casement Report. The government of Belgium, unable to simply ignore the report and with no other takers for the territory, reluctantly agreed in 1908 to take full responsibility for the Congo Free State, which was renamed the Belgian Congo, and to apply Belgian law to it. Leopold II died the following year, but not before destroying most of his archives relating to the Congo Free State.

Meanwhile, similar atrocities were being perpetrated in the Amazon basin, in particular in an area around the upper reaches of the Putumayo River, which flows into the Amazon River. In the late 19th century, control of this area was disputed between Peru and Colombia (today, the Putumayo River marks the border between the two countries). A bit to the Southwest is Iquitos, a Peruvian city on the Amazon River which can be reached by ocean-going vessels. At the very beginning of the 20th century, a Peruvian by the name Julio Cesar Arana began to buy *estancias* (local ranches, or estates) in the Putumayo area, established his own military force of mercenaries, and subjugated the local Indians into virtual slavery to force them to collect rubber from the local *Castilla elastica* trees. Although inferior to Para rubber, due to the strong demand for rubber at the time this rubber was eagerly bought by English merchants and shipped by British ships mostly to England for use by its industry. Within a few years, Arana became a very rich man.

Although both Peru and Colombia had abolished slavery by that time, the Putumayo region was beyond the control of either country's police force. Any resistance to Arana's thugs by the local Amerindians was met with severe retaliation, including large-scale murders and rapes. English and American visitors who witnessed these atrocities and reported them to their consulates were basically ignored. One young American who was deeply affected by the evils he witnessed, the engineer Walt Herdenburg, decided, after both American and British diplomatic representatives in Peru ignored his concerns, to go to England to try to raise the interest of the press. There, with the help from a local antislavery organization, he managed to get the newspaper *Truth* to publish an article on the subject. The article received a lot of attention from the public and the English Foreign Office decided to send a representative to investigate, on the grounds that many of the people working for Arana's organization were British subjects and the rubber was being sold in England.

The person put in charge of the investigation was no other than Roger Casement, the person who had earlier investigated the atrocities in the Congo Free State for the British government, and who by then was serving as the British Consul-General in Rio de Janeiro. Casement went to the Putumayo area to investigate. He delivered his report to the British Foreign Service in 1911, which sent the report to the House of Commons in February 1912. The report, executed in Casement's customary precise, yet very illuminating, manner, detailed a long list of atrocities perpetrated by Arana's men. Casement estimated in his report that out of a local Indian population of 50,000 people, 30,000 had been killed during the ten-year period in which Arana's company was operating. Casement was knighted for his investigation of the Putumayo affair, but Arana was never punished and neither were the people working for him. At any rate, by 1912 the rubber boom in the Amazon and elsewhere was reaching its end, and by 1914 Arana's company simply ceased operation.[22]

The reason the worldwide rubber price boom came to an end around 1910 was that at this point production on Para rubber plantations, mostly in Far East countries such as the British colonies in India, Malaya (today Malaysia) and Ceylon (Sri Lanka), the Dutch colonies in the East Indies (Indonesia) and the French colonies in Indochina (Vietnam, Cambodia and Laos), increased dramatically. *Hevea barsiliensis* seeds had been taken out of Brazil multiple times since the mid-19th century, with the most celebrated case being the 70,000 seeds that the Englishman Henry Wickham brought in 1876 to Kew Gardens. There was in fact no Brazilian law against taking Para rubber tree seeds out of the country, but the problem was learning how to properly grow these trees outside the Amazon, and having the economic incentives to do so. But with the help of professional botanists, it was soon discovered that Para rubber trees thrived in many tropical countries outside the Amazon basin.

When demand for rubber intensified at the end of the 19th century, the economic incentive finally appeared, and rubber trees were planted extensively in the colonies mentioned above and elsewhere. And as rubber production caught up and began to exceed demand, which happened about 1910–1912, rubber prices (but not rubber production) collapsed. At this point it was the wild trees in the Amazons that were given up, since many were old and their latex yield was low, and they were far apart from each other, making harvesting less economical. In addition, the Para rubber trees in their native habitat have a native pest, a fungus that causes the South American leaf blight, and as tappers move from one tree to another, they spread this disease, which eventually kills the tree.[23] While the rubber trees on the plantations outside Central and South America are also susceptible to this fungus, to this day the fungus has not yet been found outside its original range. Because of this fungus, rubber plantations in the Americas have never been successful. Today, the top five rubber-producing countries are Thailand (which was never colonized by Europeans), Indonesia, Malaysia, India, and Vietnam.[24] Significant rubber production also occurs in other Far East countries as well as in several African countries such as Liberia and the Ivory Coast.

RUBBER IN MODERN WARS

There are four ingredients essential for modern, mechanized war: grains, steel, oil and rubber. Of these four, rubber production has the strictest geographical distribution. Germany felt the implication of this limitation already in World War I, when it was forced to develop synthetic rubber after being cut off from natural rubber supplies. By the 1930s, 90% of all natural rubber was coming from plantations in the Far East, and during World War II both sides, the Allies and the Axis (with the exception of Japan), were cut off from most natural rubber sources. While the Japanese overran the Far East countries with the rubber plantations, except for India and Ceylon, they actually did not exploit the rubber plantations in the countries they occupied. Instead, they preferred to use the plantations' workforce as slaves in their war efforts, such as the building of the Burma railway. For the most part, the plantations lay fallow during the war.

Both in the United States and in Germany, programs to find substitutes for plantation rubber were launched during World War II.[25] For Americans, even though suitable locations for Para rubber tree plantations could be found, the immediate need for rubber precluded such a solution as newly planted trees would not begin yielding rubber for several years. Other plants with a shorter life cycle that could yield rubber, such as guayule (*Parthenium argentatum*), were considered, but the time required to adapt existing agricultural practices and land preparation for cultivation of this new crop proved lengthy, and at the end this plant did not contribute to rubber production during the war. Rubber production from the Amazon basin, accessible to American forces, also proved insignificant. Instead, the American strategy relied on three components: stockpiling natural rubber supplies as much as possible before the war (which the United States only entered at the end of 1941), limiting the use of rubber for civilian purposes, and making synthetic rubber.

During the war, 90% of the rubber produced in the United States was synthetic, and most of it was a polymer called Buna S (a World War I German invention) in which two types of molecules, butadiene and styrene, are linked together and interspersed in a long chain.[26] Styrene had to be obtained as a distillate of petroleum, but butadiene could be obtained either from a process using a petroleum fraction or from a process using grain-derived ethanol. While the butadiene made from plant ethanol was five times as expensive as that made from petroleum, the majority of butadiene going into Buna S rubber production in the United States during the war was actually made from plant ethanol, as the industrial process was already in place and no additional production facilities needed to be built. Only toward the end of the war, as newly build chemical facilities came on line, did petroleum-derived butadiene predominate in synthetic rubber production.

Germany during World War II had even less access to natural rubber. Consequently, most of the rubber made there during the war was synthetic and used petroleum-derived chemicals. A large Buna S production facility was built within the Auschwitz concentration/extermination camp complex in upper Silesia. The facility was run by the chemical company IG Farben and employed many concentration camp inmates as slave workers. Most of these inmates did not survive long and either died on the job from being overworked and underfed (as the Germans intended), or, if deemed too weak to work, by being sent to the gas chambers. A group of IG Farber officials, including Dr. Otto Ambros, the head of the Buna S factory at Auschwitz, were tried for war crimes after the war. Some were convicted, and sentenced to relatively short prison sentences. Ambros was sentenced to eight years in prison but served only six, afterward returning to professional work in Germany and later also in the US. Also near the Auschwitz camp complex, at the village of Rajsko, the Germans built a large complex of greenhouses to grow Russian dandelions (*Taraxacum kok-saghyz*), a small weed which, as the Russian scientists discovered in 1932, makes rubber of excellent quality. The greenhouse facility was run by the German government's Department of Agriculture, with the involvement of scientists from the Kaiser Wilhelm Institute, and it employed women inmates, most of whom were exterminated by the end of the war.

CONCLUSION

Previous chapters introduced plants that are beneficial to people in one way or another, and how humans have employed violent means against other people to obtain access to the lands where these plants grow, to possess these plants themselves, and to prevent other people from having them. During history, wood and rubber have fallen into the category of fought-over plant materials. Some examples of violent struggle over wood and rubber have already been presented earlier in this chapter. Another example was the ferocious struggle over the Greek city of Amphipolis between Athens and Sparta during the Peloponnesian War (430–404 BC), a struggle that took place because the city controlled the highly forested region of Thessaly in northern Greece, with its oak and pine trees essential for trireme construction. And

2,500 years later, the English and the French tried to hang on to their colonies in the Far East after World War II in part because of the rubber plantations, precipitating respectively the bloody struggles for liberation in Malaya (touched on in Chapter 2) as well as in Vietnam, a war that eventually sucked in the United States too and then spread to Cambodia and Laos.

But besides the human lives lost in the fighting over control of the sources of wood and rubber, both these materials have been used directly as components in lethal weapons as well as in equipment essential for logistic support of war, and have thus contributed to many more deaths. Wood in particular has been essential throughout our long history for constructions of weapons, and rubber has now joined wood in similar capacity. This is so because wood and rubber have similar physical properties that make them ideal for such a role: hardiness and elasticity.

Both can kill directly by making impact with the human target – the hardiness of wood and rubber is the reason that a wooden arrow or a rubber bullet can cause lethal damage when they collide with the relatively soft human body. We know that countless people have died throughout history from arrows and other wooden projectiles. But we might not be aware that rubber bullets (most are actually rubber-coated metal bullets), marketed to security forces as "non-lethal" alternatives, kill and maim hundreds of people every year around the world. They have been used heavily, for example, by Israeli security forces in the Occupied Territory of the West Bank, often with lethal consequences.

The physical strength and elasticity of wood and rubber, on the other hand, are the properties that led humans to make use of these materials in the construction the non-projectile parts of weapons as well as in machines that provide logistical support for war, and as such they are responsible for much more death and destruction then in their use as projectiles.

Their common physical properties enumerated above are derived from the same underlying principle. Both are polymers of organic molecules, and they possess remarkable strength under compressing, stretching, and shear forces to keep their physical shape: rubber does so with a bit of help from the vulcanization process. This property is achieved using the same chemical mechanism – cross-linking long linear polymers (cellulose in the case of wood, polyisoprene in the case of rubber). In the case of wood, the cross-linking molecule itself – lignin – is a polymer, and the resulting composite polymer has exceptional strength. Nature again has provided, and humans have taken advantage of this bounty that the plant kingdom has to offer, for good and bad.

NOTES

1. Levetin, E. and McMahon, K. 2012. *Plants and Society*, 6th ed. McGraw-Hill.
2. The presence of the terpene compound santalene in the heartwood of sandalwood is responsible for the characteristic smell of this wood. Sandalwood was highly prized in China for its smell, which led to overharvesting of sandalwood trees in lands around the Pacific Ocean and on Pacific Ocean islands such as Hawaii. Many other woods, such as cedar, have characteristic smells, which come from the natural preservatives found in them.
3. Vanholme, R., et al. 2010. Lignin biosynthesis and structure. *Plant Physiology* 153: 895–905.

4. Fowler, W. 1999. *Ancient Weapons*. Anness Publishing.
5. Joesephus, *The Complete Works*, translated by W. Whiston. Thomas Nelson, 1998.
6. A government-commissioned pirate was called a privateer.
7. Often, most of the casualties came from drowning.
8. Keegan, J. 1993. *A History of Warfare*. Vintage Books.
9. Logan, W. B. 2005. *Oak: The Frame of Civilization*. W. W. Morton & Company.
10. Kenyon, J. P. 1978. *Stuart England*. Penguin Books.
11. This tree, a North American native, could grow to a height of 80 meters and have a diameter of 2.5 meters at the base.
12. The shipworm has symbiotic bacteria that provide the enzymes to digest the cellulose, which the mollusk itself cannot do.
13. Rubber is short for India rubber. "India" because it came from the Americas, which Columbus mistook for India, and "rubber" because of the ability of this material to rub off pencil markings, the discovery of which is often wrongly attributed to the English chemist Joseph Priestly. Note that in other languages, the material has names that are unrelated to this ability (e.g., caoutchouc in French and gummy in German).
14. For a comprehensive description of rubber and its uses, see Tully, J. 2011. *The Devil's Milk: A Social History of Rubber*. Monthly Review Press.
15. Le Couteur, P. and Burreson, J. 2003. *Napoleon's Buttons*. Jeremy P. Tarcher/Penguin.
16. Hagel, J. M., Yeung, E. C., and Facchini, P. J. 2008. Got milk? The secret life of laticifers. *Trends in Plant Science* 13: 631–639.
17. The proteins in the rubber particles in the Para rubber tree latex are responsible for the strong allergic reaction that many people experience when they come in contact with natural rubber products.
18. Undersea telegraph cables laid down starting from the middle of the 19th century were crucial for European imperial powers in establishing control over their overseas colonies. Undersea telegraph cables also played an important role in communications between the European expeditionary forces in China and their headquarters in Europe during the Second Opium War (Chapter 5), although at that time only some segments existed, not a complete telegraph line stretching all the way to China. The full line was completed by 1870.
19. Van Beilin, J. B. and Poirier, Y. 2007. Establishment of new crops for the production of natural rubber. *Trends in Biotechnology* 25: 522–529.
20. Between 1860–1910, 60% of world rubber production came from the Amazon.
21. The eastern parts of the Congo were often raided by Arab slave traders. Part of King Leopold II's pitch to obtain control of the Congo was his professed desire to protect the Congolese from slavery.
22. Arana himself moved to England and was living there when, in 1916, the English government executed Roger Casement (not before they de-knighted him) for his role in the Irish 1916 Easter Rising. Casement, of English-Irish descent, evidently saw enough of the workings of empires during his diplomatic service to convert to the cause of Irish nationalism. Arara himself eventually moved back to Peru, where he served as a senator in the Peruvian legislature and died in 1952.
23. Lieberei, R. 2007. South American leaf blight of the rubber tree (*Havea* ssp.): New steps in plant domestication using physiological features and molecular markers. *Annals of Botany* 100: 1125–1142.
24. Total world production of natural rubber was 12 million tons in 2014.
25. Wendt, P. 1947. The control of rubber in World War II. *The Southern Economic Journal* 3: 203–227.
26. Buna S stands for Bu[tadiene]Na[the chemical symbol of the element sodium] – S[tyrene].

7 Modern Land Grabs – Hawaii, Palestine, and Latin America

FARMLAND OWNERSHIP AND NATIONAL IDENTITY

Until very recently, land for agriculture – for growing crops and grazing animals – was considered essential for the independence of nations. Most people lived on the land and engaged in agriculture, and, by and large, countries produced and consumed their own food. As nations' populations grew, they attempted to increase the land under their control by force, either to be able to have their own citizens grow the crops and graze the herds[1] or at least to have physical control over the land where their food was being produced, as we have seen for the Athenian and Roman Empires. While throughout history there have been cases of herd-tending nomads, such as the Mongols, who habitually engaged in raiding their settled neighbors, these groups either eventually settled down in defined territories themselves or ultimately did not survive as a nation.

The present-day levels of globalization of commerce in all goods, including foodstuffs, have led to the emergence of some highly urbanized countries – Singapore is an example – that can grow only a fraction of their food in their own territories but have no control over other territories. Nonetheless, these and all other countries are still concerned about being able to grow enough food on their own land as a precaution in case global commerce is disrupted, as often happens during war, so the value of land is still paramount in the affairs of nations. However, land acquisition by outright war, a method that was common until very recently, is now considered contrary to international law, a political development that has helped slow, although not completely abolish, this activity. Land sales and grants, transactions that were also once common between absolute rulers and also occurred infrequently between countries that had representative governments, have also practically ceased.[2]

The global interconnectedness of commercial and social systems in the modern world, based largely on more efficient communication and transportation technologies, have also had a major effect on land ownership. Land in one country can now be peacefully controlled by foreigners by directly buying or leasing it, or indirectly by using agents who are citizens of the country to function as "front men." Large-scale land acquisition by foreign non-state groups or even individuals, ostensibly for agricultural purposes only, has sometimes had drastic consequences on the future of the countries involved. While such consequences may have been felt purely in the social and political realms, in some cases they constituted the cause of future conflicts that ended up in war and much bloodshed. In this chapter, we will see how sugarcane cultivation led to Hawaii being somewhat (but not completely) peacefully annexed to the United States of America, how banana cultivation caused much strife in Central America, and how land purchases in Palestine for Jewish agricultural settlements, with oranges being a major crop cultivated on the newly purchased land, played a key role in the successful Zionist campaign to establish a Jewish state.

SUGARCANE AND HAWAII

In 1804, after a long military campaign, the Hawaiian warlord Kamehameha the Great (Kamehameha I) finally succeeded in bringing all the Hawaiian Islands under his control and establishing the independent Hawaii Kingdom.[3] Land in this monarchy belonged to the king, but the nobility (ali'i) were given unwritten land leases, which they parceled out to the common people under their control. In 1848, American missionaries and business people and other Westerners living in Hawaii managed to convince Kamehameha III to issue a major land reform, called the Great Mahele. Under this reform, the king retained one-third of Hawaiian land, the ali'i could buy the land that they occupied, up to a total of one-third of the land, and assume legal title to it, and the peasants could each own three acres, also up to an aggregate total of one-third of the land. Furthermore, from then on land could be bought and sold to anyone, including to foreigners.

Many of the ali'i were at that time highly in debt to Westerners, and the peasants did not quite understand the concept of land ownership and often did not have the knowledge to claim ownership or the money to pay the tax on the land. As a consequence, westerners soon owned 80% of all privately held land in Hawaii (i.e., land not under the direct control of the monarch). As it happens, the land reform coincided with the decline of the worldwide whaling industry, affecting Hawaii since it served as a major hub and resupply center for whaling ships in the Pacific. The land reform also coincided with the Gold Rush in California, to where millions of people were moving to dig for gold, not to engage in agriculture. The gold miners needed to be fed, and growing agricultural crops in Hawaii and shipping the produce to California was more cost-effective than sending food overland from the eastern parts of the United States. Suddenly, cash crop agriculture became Hawaii's major business, one for which the locals had no affinity nor the necessary capital and connections, but for which the Americans in Hawaii were perfectly suited. In particular, the descendants of the missionaries, some of whom married local ali'i women and therefore came into possession of large tracts of land, formed a politically well-connected (and genetically related, since they intermarried among themselves) network of plantation owners and businessmen. This group eventually expanded into banking and shipping, forming the five large companies, dubbed The Big Five, that controlled most business activities on the islands.

Soon, sugarcane (Chapter 3) became the biggest cash crop industry in Hawaii.[4] Sugarcane had been one of the plant species that Polynesian seafarers carried with them on their colonizing voyages, and it was therefore already growing on the Hawaiian Islands by the time the Europeans and Americans arrived. In fact, the first sugarcane planation was established in Koloa, Kauai, in 1835. But the Gold Rush in California and the attendant increase in population on the west coast of the United States was a major factor in encouraging the growth of sugarcane plantations in Hawaii. With the onset of the Civil War in the United States in 1860, the Union states could no longer buy sugar from the Confederate states in the South and sugar prices went up considerably, further stimulating the growth in sugarcane cultivation in Hawaii. Finally, the completion of the Transcontinental Railway in 1869 and the resulting decrease in transportation costs made Hawaiian sugar even more competitive – pricewise – throughout the United States. And to make the business highly profitable, plantation owners looked for workers that were willing to work very hard for very little pay. They obtained such workers by importing indentured servants to Hawaii from China, Japan, the Philippines, the Canary Islands, and several other countries.

With the sugarcane industry being the major engine of economic activity on the Islands, the Americans running this industry demanded ever more political control over how the Kingdom of Hawaii was managed. In 1887, King Kalakaua gave in to this pressure and approved a new constitution (dubbed the "Bayonet Constitution" by the natives for the way it was imposed on the Hawaiian monarchy) that transferred more power from the king to the legislature, which was dominated by Americans and others of European descent. However, when King Kalakaua died in 1891 and Queen Lili'uokalani ascended to the throne, she revoked the 1887 constitution and removed the haole (a Hawaiian word for non-Hawaiian people) from her cabinet.

It was in the interest of the sugar industry in Hawaii, and in the interest of The Big Five, to have the kingdom annexed by the United States, because if Hawaii became a US territory there would be no tariffs on sugar and other Hawaiian commodities imported into the United States. The Americans in Hawaii, led by Lorrin Thurston, a grandson of pioneering missionaries on both his father's and his mother's sides, used the cancellation of the 1887 constitution as a pretext to form a group first called The Annexation Party, and later The Committee of Safety. Thurston then traveled to Washington DC and met secretly with President Benjamin Harrison. The president assured Thurston that he was in favor of annexation. When Thurston returned to Hawaii, the Americans in the legislature embarked on a strategy of creating a parliamentary crisis by refusing to pass a budget, hoping that this would lead to economic problems and subsequent American intervention. But before this obstructionist plan could succeed, the Committee of Safety heard that the Queen was about to announce a new constitution in which haole would have no political power, and they decided to act right away. In January 1893, backed up by American marines from the ship the USS Boston, which was docked at Honolulu's Pearl Harbor port, they declared themselves a provisional government. Queen Lili'uokalani, wishing to avoid bloodshed, resigned under protest.

Next, Thurston and the Committee of Safety traveled to Washington DC to ask the new president, Grover Cleveland, to annex Hawaii as a territory. Queen Lili'uokalani also traveled to Washington DC to argue her case and to convince the president to help her restore the Hawaiian monarchy and Hawaiian independence. After hearing both sides, Cleveland then sent his own representative to Hawaii to investigate. Based on the report received from this emissary, the president denounced the American revolutionaries and the involvement of the American Armed Forces in it, refused to support the annexation of Hawaii, and vowed to help the Queen restore the independent kingdom of Hawaii. However, the local Americans continued to control Hawaii by force and in 1894 declared the Independent Republic of Hawaii. In 1897, a new president, William McKinley, took office. Soon afterward, the United States fought and won the Spanish–American War, and the Spanish colony of the Philippines, on the other side of the Pacific Ocean, became an American possession (as well as Guam). Making Hawaii, situated in the middle of the Pacific Ocean on the way to the Philippines, part of the United States now made good strategic sense, and by an act of Congress, signed by President McKinley, Hawaii became an American Territory on July 7, 1898. From then on, Hawaii served as a major base for US military operations in the Pacific, which is why it was massively attacked by the Japanese in World War II. Hawaii became a state in the United States in 1959.

ORANGES AND ZIONISM IN PALESTINE

In 1848, a wave of revolutions swept Europe. There were many factors leading to them, somewhat different in each country, but they were all essentially democratic in nature. The revolutionaries were rebelling against autocratic and imperial regimes, demanding more rights and improved conditions for the general populations. While none of these 1848 revolutions were successful in permanently unseating the existing regimes, they did have major effects on future historical developments of Europe. For one thing, the common people did eventually receive additional individual rights from their rulers. For another, the sentiment of nationalism, involving a group identity based on perceived shared blood, land and history, was firmly launched. This is why this period has been called the Spring of Nations. In fact, the heightened feelings of nationalism led to a general competition in Europe, with each nation trying to enhance its economic status and its prestige by various means. One of them was a competition for overseas colonies. Since the New World was by then completely colonized by Europeans, the new impetus for colonization was by necessity focused on Africa as well as parts of Asia.

The rise of nationalism in Europe had complex but often negative effects on national minorities living within the boundaries of each state. One such minority was the Jews, a religious and ethnic group that traced its origin two thousand years back to an area in the Middle East that was often referred to as Palestine (after Palaestina, a name that the Romans coined). In the 19th century, populations of Jews were living in every European country (and many other places), but the majority of European Jews were to be found in Eastern Europe, and most of them within the borders of the Russian Empire, which then included large parts of Poland and Ukraine. While Jews had always experienced discrimination in Europe based on their religion, the rise of nationalism also gave rise to racial antisemitism. Outside Eastern Europe, antisemitism was generally expressed in covert and overt discrimination that did not cross the line into violence. However, in East Europe, and particularly within the Russian Empire, the Jews were often officially persecuted by the authorities and subjected to "pogroms," governmentally instigated and sanctioned massacres. Pogroms became particularly prevalent after the assassination of Tsar Alexander II in 1881.

The dire situation of the Jews in Russia led more than two million of them to emigrate between 1880–1920. Most of these migrants went to the United States and other countries in the New World. However, clearly influenced by the ascendance of the ideology of nationalism in Europe, some Russian Jews considered the possibility of reconstituting the Jews as an independent nation. They were eventually supported in this notion by some other European Jews who were living in more tolerant countries but nonetheless were experiencing serious antisemitism that led them to doubt that Jewish assimilation could be successful in Europe. Foremost among them was Theodore Herzl, a Hungarian secular Jew who, as a correspondent in Paris for an Austrian newspaper, covered the famous Dreyfus Affair.[5]

While several possibilities for a territory for the reconstituted Jewish nation were considered at one point or another, Palestine was always the main contender from the beginning of the Zionist movement. Indeed, the term Zionism connotes the ideology that calls for the establishment of a Jewish homeland in Zion, the Biblical name of Palestine, and after a few internal debates within the budding Zionist movement, Palestine was soon accepted as the place where the Jewish homeland should be rebuilt. But the early Zionists faced several obstacles in accomplishing this goal. Palestine in the 19th century was part of the Turkish Ottoman Empire, and it was inhabited, although somewhat sparsely, by a mostly Arab population.[6] In 1881, there were 457,000 inhabitants in Palestine, 13,000–25,000 of them were Jewish and almost all the rest were of Arab ethnicity (the vast majority of them Muslim). Members of the small Jewish minority in Palestine lived predominantly in four cities – Jerusalem, Hebron, Safad, and Tiberias – and did not engage in agriculture.[7] And the European Zionists were not in control of a state that could simply order its army to invade Palestine and take possession of it. In the face of these limitations, the Zionist plan of action that evolved, first by multiple, separate initiatives and eventually by a coordinated effort, was simple: buy agricultural land in Palestine and settle (mostly) eastern European Jews on it.

Palestine is a relatively small area, approximately 400 km in length from north near the sources of the Jordan River to Eilat on the Red Sea in the south, and 125 km at its widest from the Mediterranean Sea to the Syrian-African Rift Valley.[8] The southern half is a desert, and the northern part is mostly hilly, with narrow plains by the coast and on the eastern border along the Jordan River, and another plain, called the Jesre'el Valley, extending from Haifa on the Mediterranean coast south-east to the Jordan River. In 1882, when the first permanent agricultural settlement of Jews in Palestine was established, 75% of the population in Palestine consisted of farmers, and most of the farmers lived in the hilly areas. The valleys, and in particular the coastal plain, were not heavily populated for several reasons. One was security – the Turkish administration was not able to firmly secure the country and robberies by roving armed Bedouin tribes were common, particularly in the flat plains where defense was more difficult. At least equally important, the mouths of the local streams that carried water from the hills to the Mediterranean Sea during the rainy winter season were often blocked by sand dunes, causing swamps to be formed in the plain (Figure 7.1). The swamps were breeding grounds for the mosquitoes that carried the malaria parasite, and the coastal area was therefore a place where malaria was prevalent. There were swamps and malaria-carrying mosquitoes in some of the inland valleys in the north as well.

For historical reasons too complex to explain here in detail, when the Zionist movement was launched in late 19th century, most of the arable lands, particularly in

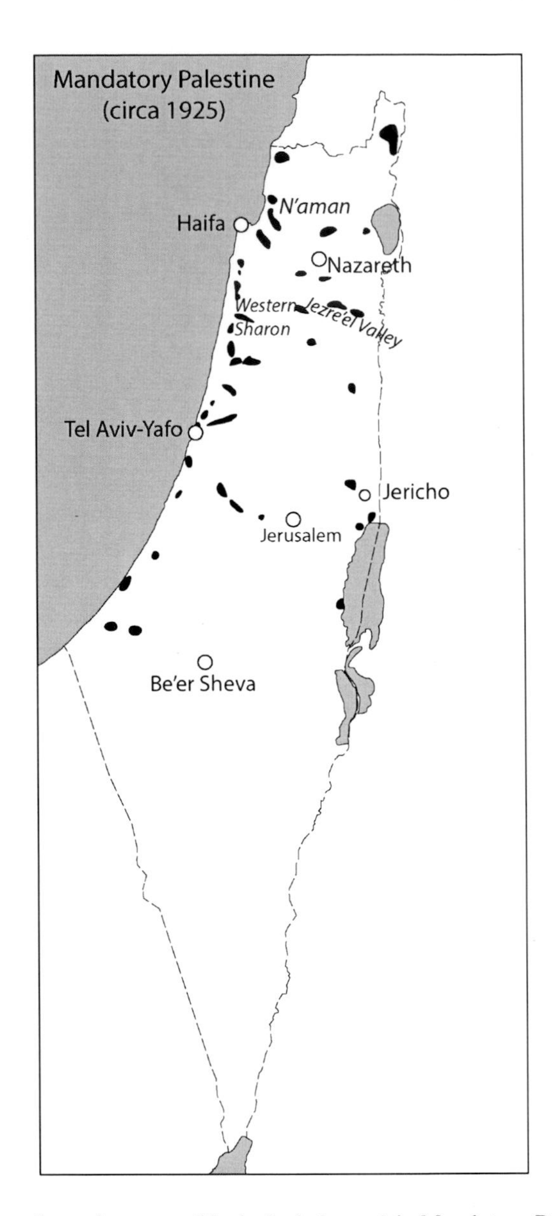

FIGURE 7.1 Locations of swamps (black-shaded areas) in Mandatory Palestine c.1925[29].

the plains, was in the hands of absentee landlords, rich Arabs who lived in the major cities of Palestine as well as in Damascus, Beirut, and even beyond. The land itself was cultivated by tenant farmers whose families had lived on the land for hundreds of years. In the late 1800s, Jews began buying land from these absentee owners, initially raising money from rich Jewish donors in Europe such as Baron Rothschild of France and Baron de Hirsch of Bavaria. In 1901, the Jewish National Fund (JNF) was established by the Fifth Zionist Congress in Basel[9] and tasked with buying agricultural land in Palestine and helping to settle Jews on it. The JNF raised the money to accomplish this goal by soliciting both small and large donations in most Jewish communities in the diaspora. From the beginning, the Jewish land purchasers were willing to pay high prices for the land, since the purpose of obtaining the land was more political than economic. The calculations of return on investment was not paramount, particularly since the people donating the money did not need or expect to make any profit at all.[10] The high offers for the land often induced the absentee Arab owners to sell.[11]

Once the new Jewish owners took possession of the land and established their settlements, the Arab tenant farmers were evicted from the land and given meager, one-time monetary compensation. While the first Jewish settlement, Zikhron Ya'akov, was established in 1882 on a hill south of Haifa and, after a few years, specialized in growing grapes, most of the subsequent Jewish land purchases and settlements took place in the low-lying, often swampy, and generally less-populated areas. The financial backing for the Zionist settlement efforts by European and American Jewish organizations and wealthy individuals meant that the settlers could bring resources and mechanical equipment to improve the land. Such improvements included constructing access roads, draining the swamps by digging ditches and opening the blocked streams, and installing irrigation systems with mechanical water pumps and a network of water pipes.

To keep the drained swamplands dry, the Jewish settlers introduced a eucalyptus tree species native to Australia, *Eucalyptus camaldulensis* (Figure 7.2). This tree grows along river beds in Australia and has a very high transpiration rate. In Palestine, it was planted in small groves and along plot lines in the former swampy areas, and it helped lower the water table further. Since these trees grow fast, they could be cut down to the ground (an agricultural practice called coppicing) to generate new growth, and the trees became a good source for wood used for construction of fences and other buildings and for firewood.

FIGURE 7.2 Left, a tree branch with flowers of *Eucalyptus camaldulensis*, with mature fruit shown in top left corner. Right, a branch with flowers of an orange tree, with the fruit shown below.

The coastal lands in Palestine are relatively light, and in some places quite sandy. A crop that does well in these soils is citrus. *Citrus* is a plant genus that includes oranges, lemons, limes, grapefruits, pummelos, mandarins, and tangerines. Many of the presently cultivated citrus fruit trees are hybrids between various species in the genus, and the initial domestication of citrus trees is believed to have occurred in South Asia thousands of years ago. However, recent studies suggest that the ancestral citrus lineage originated in Australia, New Caledonia, and New Guinea.[12]

Today, citrus fruits in the aggregate produce the second biggest fruit crop in the world (after bananas, more to follow), with about 115 million tons harvested in 2007. Oranges account for half of this total, and while today most oranges are used by companies to make pasteurized juice in an industrial process – modern consumers prefer to enjoy the fruit without getting their hands dirty by peeling it – this is a phenomenon that only became commercially significant in the second half of the 20th century. Until then, most citrus fruits were first peeled, then the sections separated and eaten.

Orange fruit contains a fair amount of many vitamins and minerals that the human body needs. Compared with other food sources, oranges and other citrus fruits are particularly rich in vitamin C (ascorbic acid), a compound that humans cannot synthesize themselves but which is essential to metabolic processes in the human body. Humans must obtain vitamin C from plants, and the lack of vitamin C leads to the devastating disease scurvy. Scurvy was famous for afflicting sailors on long-term voyages during the Age of Discovery, before the cause of scurvy was discovered and preventive steps such as the consumption of fresh citrus fruits could be taken (see Note 4 in Chapter 4).

Oranges are also rich in the sugars glucose, fructose, and sucrose, and it has a complex flavor that depends, in addition to the aforementioned sugars, on the presence of several organic acids, pectin (which imparts a specific "mouth-feel" sensation), and as many as 30 different volatiles.[13] These volatiles fall into several classes of chemicals, including alcohols, aldehydes, ketones, hydrocarbons, and esters, all of which are derived from either the fatty acid or the terpene biosynthetic pathways. In fact, unlike grapefruit flavor, which is dominated by two main volatiles, the sesquiterpene nootkatone and the sulfur-containing compound ρ-menthen-8-thiol, orange flavor is not dominated by one or a few volatiles. However, orange fruit peel is extremely rich in the monoterpene limonene, which is therefore found in relatively high levels in commercially prepared orange juice, since the whole fruit, with the rind, is squeezed in the industrial process.[14]

While citrus trees originated in the tropics, they do well in low altitude, subtropical regions with moderate climates, and they do particularly well along the coast of most Mediterranean countries (Figure 7.3). Some citrus trees were brought to the Mediterranean region from China and India by Arab traders before the 11th century, but sweet oranges (*Citrus sinensis*) were brought to Southern Europe from China only at the beginning of the 16th century by the Portuguese, and from there these oranges soon spread to the Middle East. Cultivated citrus trees are mostly propagated vegetatively, by cutting a shoot ("scion") and grafting it onto a rootstock.

In Palestine in the second half of the 19th century, the local Arab farmers developed a successful orange variety called Shamouti (later named Jaffa Orange, because of the port from which the oranges were shipped to Europe). The fruit is

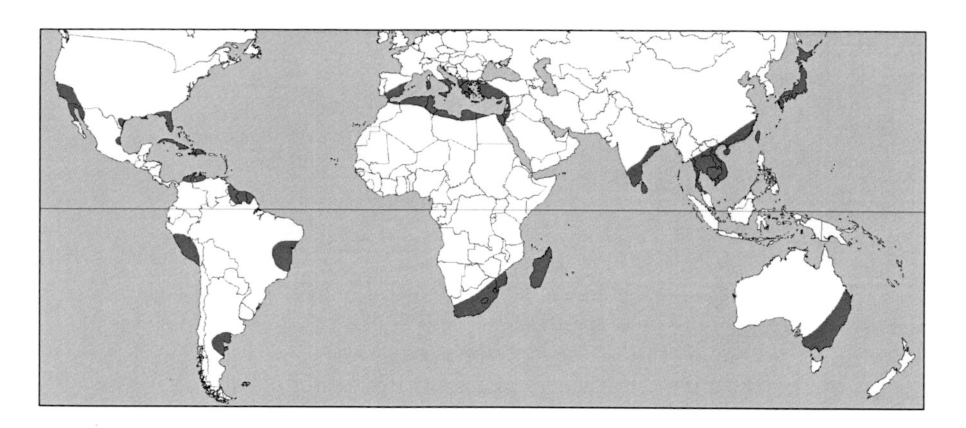

FIGURE 7.3 Areas around the world where commercial citrus orchards are currently cultivated (indicated in dark gray).

large, practically seedless, and it has a thick rind that makes the orange easy to peel. It was prized for its size, flavor, and long shelf-life, keeping its freshness for weeks without refrigeration after being harvested.[15] With the widespread adoption of steam ships, oranges could be shipped from Palestine to northern Europe in non-refrigerated ships in just a few days. The harvesting season of the oranges extended (and still does) from October through Christmas and beyond, a time when few fruits are available in Europe. This long harvesting period is due to the durability of the fruit which, once it ripens, stays on the tree and does not drop or spoil. Therefore, there was no need to employ an especially large work force for a short period of time or to use mechanized equipment to pick the fruit (both economically expensive alternatives), making orange cultivation ideal for the then relatively unsophisticated agricultural economy of Palestine. Thus, in the late 19th century, growing oranges for export became the most economically important agricultural activity in Palestine, whose population was still overwhelmingly rural and whose economy was mostly agrarian, supplanting cotton and bringing more foreign revenues into the economy than any other segment of the economy.

The orange groves were concentrated in the coastal area and the oranges were shipped from Jaffa, an ancient port city in the center of the country, with the majority of the fruit going to England. The orange industry was typically controlled by Arab city merchants who owned the land and had the capital to invest in planting and maintaining the trees until they bore fruit. The groves employed many workers who tended the trees, watered them, and picked the fruit when ripe. Additional employment was provided in packing the fruit, transporting it to the port, and loading it on the ships.

Jewish Zionist immigration to Palestine, known as "Aliya," came in several recognized phases, or waves. The first wave of immigrants, who began arriving, as we saw, in 1882, did not pick up orange cultivation as an occupation right away, but soon the Jewish settlements in the valleys, and particularly on the coast plains, began growing oranges. However, they often employed Arab workers to tend the trees. In 1903 the second Aliya wave began, and this time most of the Jewish immigrants were committed socialists from Eastern Europe (but nevertheless also committed nationalists) who believed in working with their own hands. Starting from this second Aliya wave, larger than the first, a struggle over land and work developed in Palestine between the Jewish community and the Arab community. While this struggle was quite complex and involved many aspects of the economy, both in agriculture and in urban industry, the citrus groves played a central role in it.

As already described, the leaders of the Zionist movement understood very well that national rights have always been accorded to those who live on and work the land. Therefore, the settlements and the agricultural activity that the Jewish settlers engaged in were essential for achieving Zionism's political goal of establishing a Jewish state in Palestine. These efforts yielded results, first slowly but then, after the British conquest of Palestine in 1918, the pace of Jewish immigration to, and the agricultural settlement of, Palestine accelerated. The increase in immigration under British rule was due mainly to the announcement in November 1917 by Lord Balfour, the British Foreign Secretary (in a letter to Baron Rothschild, a leader of the Jewish community in England), that England would act to support the establishment of a "Jewish National Home" in the country. The Balfour Declaration was made because of strong lobbying efforts on the part of the representatives of the Zionist movement in England, foremost of them the chemist Dr. Chaim Weizmann, and it was taken without consulting the Palestinian Arabs. Thus, while the Jewish population in Palestine grew from 13,000–25,000 (2.7%–5.4% of the total population) in 1881 to 60,000 in 1914 (7.5% of the total population), and then declined somewhat because of the deprivations of World War I to 59,000 in 1918 (7.9% of the total population), when the British took over control of the country from the Turks, it then began to greatly accelerate. On the eve of World War II in 1939, the Jewish population in Palestine reached 445,000, 30% of the total population.

Much of the land that Jews purchased in Palestine was planted with orange trees. In a sense, orange trees were the major, although certainly not the only, weapon that the Zionists used, in their own words, to "conquer the land" (the term was indicative of an attitude that ignored the people living on the land). Not only did orange trees grow well on sandy or generally light soil, they also provided major employment opportunities for the newly arrived Jewish immigrants. By 1929, half of the 6,000 hectares of orange groves in Palestine were owned and operated by Jews. By 1939, the total amount of acreage planted with oranges tripled to roughly 18,000 hectares, with the majority of the oranges groves in Jewish hands. Furthermore, starting with the second Aliya wave that began in 1903, Arabs were generally excluded from working in all aspects of the Jewish-owned orange industry. Indeed, the Zionists in Palestine tried, and mostly succeeded, in developing their own separate economic community that excluded Arabs from it. With oranges, this took some special efforts as the Arab merchants of Jaffa initially played a major role in buying citrus produce and arranging for its export via the Jaffa port (and later the Haifa port), but the Jews eventually established their own purchasing and exporting corporations and even developed their own port in Tel Aviv, initially as a response to the Arabs' efforts at blocking access to the Jaffa port to Jewish products during the 1936–1939 period of hostilities (more to follow).

From the beginning of the Zionist settlement activity in Palestine, many of the local Arab residents understood the objectives of this activity and naturally objected to it.[16] They realized that a Jewish state would have laws and customs that were foreign to them and may not treat them well. They therefore consistently demanded, first from the Turkish rulers and then from the British, that Jewish immigration to Palestine be stopped and that land purchases by Jews not be allowed. Their pleas were generally not successful, and as we have seen, under the British Mandate Jewish immigration, land purchases, and settlements accelerated. Eventually, the Arabs resorted to violent actions to stop the implementation of the Zionist program. Major flare-ups occurred in 1920–1921 and 1929, and in 1936 what began as a general Arab Palestinian strike evolved into an intermittent war against both British and Jewish targets (including the burning of dry, ready to be harvested wheat and barley fields, see Chapter 2) that continued until 1939. The civilian and military resistance of the Arab Palestinian population during the 1936–1939 period was later dubbed the Great Arab Rebellion. Both the British and the Jews in Palestine fought back against the rebelling Arabs, and while Jewish immigration did slow down due to some restrictions placed by the British in an effort to mollify the Arabs, the rebellion was not successful in stopping new Jewish land purchases and new settlements. World War II broke out in September 1939 and Jewish immigration and the Jewish–Arab military conflict went into a hiatus, but the conflict resumed in full force after the end of the war in 1945.

Starting first during the 1936–1939 Arab-Palestinian rebellion, and as a direct response to it, successive political efforts were made by the British and the rest of the international community to resolve the conflict by proposing the division of Palestine into two states, a Jewish one and an Arab Palestinian one. These efforts culminated in the 1947 UN Partition Resolution, which called for the creation of a Jewish state on just about 50% of the land of Mandatory Palestine, including most of the coastal area. No doubt the fact that large swaths of land – 4% of total Palestine (including the desert), but a much larger portion of the arable land in the coastal and interior plains – were already owned, settled, and cultivated by Jews was crucial in the determination of which part of the country was awarded to the Jewish state. Indeed, almost all the land already owned by the Jews, much of it planted with orange trees, was included in the area proposed for the new Jewish state. Thus, land acquisition followed by the establishment of agricultural settlements in which orange cultivation played a major role was a very important element in Zionism's success in "establishing facts on the ground" (another Zionist term). As is well known, the partition resolution was not accepted by the Arab side and this rejection led to the armed conflict of 1948–1949, in which the Palestinian Arabs lost substantially more territory to the newly established Jewish state of Israel. The Arabs living on the lands conquered by Israeli forces in the 1948–1949 war were for the most part uprooted and became refugees. In the intervening 70 years between that war and now, there have been many additional wars centered around the territory of old Palestine, and the conflict is nowhere close to a peaceful resolution.

BANANAS AND CENTRAL AMERICA

The banana "tree" – it is actually a large herbaceous plant (Figure 7.4), related to the grasses – was first domesticated as early as 10,000 years ago in New Guinea and

FIGURE 7.4 A banana tree, showing the terminal inflorescence with the young banana fruits developing (by parthenocarpy, meaning without pollination) from the female flowers, and the unopen male flowers (inside the red sheath).

probably also elsewhere in South Asia.[17] Banana cultivation soon spread throughout the tropical and subtropical parts of East Asia. The banana was one of the few staple food plants that the Polynesians took with them on their ocean voyages as they colonized the Pacific Islands. Bananas reached sub-Saharan Africa, probably by both the land route and by sea, no later than the 5th century AD. The Arabs brought bananas first to the Middle East and then to North Africa and the Iberian Peninsula. Finally, in the 16th century bananas were brought to the New World by the Portuguese, completing their encirclement of the planet.[18]

Bananas and related species comprise the genus *Musa*. The Polynesian bananas, called Fe'i bananas, are derived from the species *Musa lolodensis*, while all other domesticated banana cultivars are derived from two other species, *Musa acuminate* and *Musa balbisiana*. While there are some cultivated bananas that have large seeds, it is believed that many of the initial domesticated cultivars were diploid (having two sets of chromosomes) and parthenocarpic, meaning that the fruit starts developing without pollination, so no seeds are produced. Today, most of the banana cultivars grown for food are triploid (with three sets of chromosomes per cell). In the case of triploid cells, meiosis, the process that generates gametes with half the number of chromosomes that the cells of the parent have (see Glossary), yields gametes with 1.5 sets of chromosomes (on the average). Two such gametes, when fused, form a zygote with unbalanced numbers of chromosomes (e.g., two copies of some chromosomes, three of others, and four of yet other chromosomes), and they cannot develop into viable embryos. Therefore, the fruits of triploid banana have no seeds, even when pollination takes place. Triploid bananas have another advantage. Since triploid cells tend to be bigger in general, tripolod banana fruits are larger than those of diploid bananas. However, triploid bananas, as we will see below, must be propagated vegetatively.

The common distinction between sweet, dessert bananas that contain a fair amount of soluble sugars (mostly glucose, fructose, and sucrose) and can be eaten directly when ripe, versus starchy plantains that must be cooked to be edible is only partially predicated on taxonomy. Indeed, most dessert banana cultivars are triploids of *M. acuminata*, while plantains are triploid hybrids of *Musa acuminata* and *Musa balbisiana*. However, some *Musa acuminata* X *Musa balbisiana* hybrids can also be eaten as fresh sweet fruit when ripe. As banana and plantain fruits develop, both may contain up to 70%–80% starch (by dry weight). But in dessert bananas, the starch is almost completely degraded by amylase enzymes and converted to glucose, fructose, and sucrose during the ripening process, while in fruits of accessions that are called "plantains," except in a few cultivars as noted above, the enzymatic process of starch conversion to mono- and disaccharides does not occur.

Both banana and plantain fruits also contain small but nutritionally significant amounts of many vitamins and minerals. For many subsistence farmers, particularly in Asia and Africa, banana fruit represents a major part of their diet. Banana and plantain flavor is determined by the presence of sugars, organic acids such as citric and malic acids, and numerous volatiles.[19] A large number of these volatiles are esters in which the acid moiety comes from degradation of fatty acids or aliphatic amino acids such as leucine, and the alcohol moiety comes from degradation of fatty acids.[20] Other volatiles that contribute to banana flavor are aldehydes and alcohols derived from the degradation of fatty acids, and eugenol, which is also the dominant flavor compound in clove, as described in Chapter 4. Curiously, banana fruits contain a fair amount of the animal neurotransmitter dopamine, up to 10 mg per 100 g of fresh banana pulp (and 50 times as much in the peel).[21] Dopamine is not volatile and is not likely to affect flavor perception. It has been hypothesized to function as an antioxidant in the fruit but may actually act as a defense compound.

Today, bananas are grown in tropical and subtropical areas all over the world. The total annual tonnage of banana fruit harvested worldwide, about 140 million, makes it the most common fruit in the world. However, unlike oranges, most of the bananas grown – 87% – are consumed locally, or at least within the country of origin. Since bananas are a great source of carbohydrates and many vitamins and minerals, they have been adopted as subsistence food by many poor people around the world, including in South Asia, Africa, and Latin America, where banana cultivation has penetrated deep into the jungle. For example, the Yanomamö Indians in Brazil and Venezuela, a group of indigenous tribes that had little contact with other people until very recently, picked up banana cultivation from neighboring tribes several hundred years ago.[22] Currently, the largest producers of bananas are India (18% of world harvest) and China (8%), and many other East Asian as well as African countries grow a substantial amount of bananas, but few of the bananas grown in these countries are exported. The five largest exporters of bananas – such exports go mostly to North America and Europe – include the two Central American countries Costa Rica and Guatemala, the South American countries Colombia and Ecuador, and the Philippines in South Asia.

Again unlike oranges, bananas are highly perishable fruits, and developing bananas as a cash crop for export required a fair amount of ingenuity. In fact, their extremely low price in developed countries' grocery stores – particularly when compared on a per calorie basis with some locally grown fruits such as apples[23] – is quite astounding.

Dessert bananas were first exported to the East Coast of the United States soon after the Civil War. They came from the Caribbean Islands, the closest region to the United States where bananas were growing (although bananas could be grown in Florida and the rest of the extreme Southern parts of the United States, the climate there for banana cultivation is marginal). However, in those days sailing time from the Caribbean to the American Northeast ports of Philadelphia, New York, and Boston could take as much as three weeks, longer than the shelf-life of the fruit. Bananas, which are picked green, begin to ripen right away if kept at temperatures above 59°F; if stored below 55°F for more than a few hours, the ripening process is permanently disrupted. Even if the trip were accomplished at a faster clip, there would not be much time to distribute the fruit to local merchants, and certainly not to send it to destinations outside the port city. Consequently, bananas that did reach American ports in good shape were few, and therefore expensive. They were considered luxury items.

That fresh bananas became cheap and available all over the United States, and eventually in Europe as well, was due to two separate developments in late 19th century. In 1870, a Boston sea captain by the name of Lorenzo Dow Baker picked up some banana bunches in Jamaica on his way back from Venezuela, where he had gone to transport some gold miners. His fast ship made the trip from Jamaica back to Boston in 11 days, and he was able to make a $6,400 profit from selling his fresh bananas to local merchants. Seeing how lucrative the banana trade was, the next year he bought land in Jamaica and started growing bananas and shipping them to Boston and other East Coast ports. In 1885, he teamed up with a partner, another Bostonian by the name of Andrew Preston, and together they established a company they named Boston Fruit. It was Preston who was instrumental in developing a distribution network throughout the country that relied on railway transportation and local refrigerated storage facilities. Eventually, the sailing ships were replaced with faster steamships that were soon equipped with refrigerated cargo hulls.

The second set of events that led to the popularity of bananas in the United States was even more fortuitous. In 1871, the government of Costa Rica signed an agreement with a US railroad builder, Henry Meiggs, to build a railroad from San Jose, its capital in the center of the country, to the city of Limon on the Caribbean coast. Meiggs died in 1877, but his nephew Minor C. Keith, an engineer, continued the project. When the Costa Rican government could no longer make payments for the construction work, it instead gave Keith 800,000 tax-free acres of land along the railway tracks and a 99-year lease on operating the railway. The operation of the railway did not bring much money to Keith. However, during the construction of the railway he had planted banana groves along the railway track to feed the workers, and now he discovered that he could make a lot of money by shipping the bananas to the coast on his train, then shipping them to the United States for sale there. His company, the Tropical Trading and Transport, went on to acquire more land, and to develop banana plantations in other Central American countries as well as in Colombia.

In 1899, Boston Fruit and Tropical Trading and Transport merged to form United Fruit Company (UFC), which has also been known throughout Latin America as El Pulpo, or The Octopus (in 1970 it was renamed United Brands Company and in 1984 it was further renamed Chiquita Brands International, its current name). At some point during the early to mid-20th century, UFC controlled 80% of the banana market in the United States, with its closest market competitor being the Standard Fruit Company (today, Dole Food Company). It achieved its preeminent market share in the United States by buying out many American banana importers. But its nickname El Pulpo was based more on its behavior in the producing countries.

This nickname is partially a reference to its business model, which relied on vertical integration – controlling land, production work force, transportation, and communication systems within the producing countries (thus constructing and controlling the local railways and the radio and telephone systems), shipping to the United States, and controlling distribution there. To control production cost in Latin American countries, UFC (as well as Standard Fruit) was not squeamish about bribing politicians to gain favorable tax rates and lax regulations concerning the treatment and payments of workers. The large proportion of the general economic activity concentrated in the banana production sector in these countries, and the corruptive influence of the American banana companies on the political systems of these countries, earned them the derogative sobriquet "Banana Republics."[24]

In fact, the local dictators often helped and abetted the horrible treatment that UFC accorded its workers. A famous case is the event known as the Banana Massacre, which occurred on December 6, 1928, in Ciénaga, Colombia. Major strikes by banana plantation workers, demanding better working conditions, had begun in Colombia in October of that year. But on December 6, a Sunday, thousands of banana plantation workers and their families gathered in the Ciénaga town square to hear a speech by a regional governor. The Colombian army, however, opened fire on the people in the square, killing perhaps as many as 3,000 people.[25] A few days later, the American ambassador to Colombia, Jefferson Caffrey, sent the following telegram to Washington: "I have the honor to report that the Bogota representative of the United Fruit Company told me yesterday that the total number of strikers killed by the Colombian military exceeded one thousand." The military officer responsible for the massacre justified his action by saying that if he did not act, the US military would have invaded the country, and as a patriot he wanted to prevent an invasion of the hated Gringos.

The general antipathy in Colombia against UFC resulting from the perception that it was complicit in this massacre soon led the company to withdraw its operation from Colombia. But the country itself began sliding into a long social and political upheaval that has lasted to this day and has involved various drug cartels and guerrilla organizations controlling large parts of the country, with a huge toll of civilian casualties (see Chapter 6). Chiquita Brands eventually restarted operations in Colombia by buying bananas from local farmers, but in 2007 the company pled guilty in United States Federal Court to paying $1.7 million to a Colombian right-wing "death-squads" organization, The United Self-Defense Forces of Colombia, to kill and intimidate union organizers and banana farmers into giving Chiquita better terms (thus improving the bottom line of the company).

The main reason UFC was called El Pulpo, however, was that it kept acquiring more and more land in Latin America, using its political clout to get outright concessions from governments or to buy the land on favorable terms. The reason for this out-of-control land acquisition urge was rooted in biology – the biology of banana root disease. Soon after bananas were introduced to Central America, a root pathogen, the fungus *Fusarium oxysporum*, was discovered to infect bananas in Central American banana plantations. The fungus grows inside the xylem, the vascular system responsible for conducting water from the roots to the shoots and leaves. The growth of the fungus blocks the movement of water in the xylem vessels so that the stems and leaves wilt and the plant dies. The fungus was first formally identified in banana plantations in Panama at the beginning of the 20th century, and therefore the disease has been called "Panama Disease." However, it is now believed that the fungus originated in Asia, and in retrospect it appears that the first outbreak of this disease occurred in Australia in 1876.

At the time that Panama disease became endemic in Latin America, all bananas grown there for import to North America were of a single variety, called Gros Michel ("fat Michel" in French). This variety turned out to be highly susceptible to the wilt fungus, most likely because, given the large number of identical banana plants being grown on plantations, the fungus had an increased chance of evolving specificity and virulence to this variety. Gros Michel was actually a recent (early 19th century) introduction from Asia to the Caribbean, and it was highly favored by consumers for its flavor. Because Gros Michel is a triploid, as are most dessert bananas, it could not be crossed with other varieties that have resistance to the fungus to form a resistant hybrid, a standard plant breeding technique to make crop plants resistant to diseases. To make things worse, the fungus is resistant to all chemical treatments, and after killing the banana plants, it can live in the soil for up to 30 years. If farmers try to plant new banana plants that are not resistant to the fungus in that same plot of land, the plants will get infected and will not survive to set fruit.

In the absence of a treatment to get rid of the fungus, UFC and other banana companies embarked on a strategy of planting bananas in new, uninfected soil. It was not actually a very smart thing to do since bananas are propagated vegetatively, by breaking parts of the underground root system (the corms, or thickened rhizomes) and planting them elsewhere. As new banana plantations were established on new lands, invariably some corms carried the wilt fungus with them and the disease would soon spread to all plants in the new plantation. But the UFC and Standard Fruit companies were desperate, and they used all their political clout to obtain more land. Not surprisingly, this approach eventually failed, and in the late 1950s the companies threw in the towel and replaced Gros Michel with a new triploid banana variety, the Cavendish, which was resistant to the specific *Fusarium* strain that Gros Michel was susceptible to. However, the Cavendish has by all knowledgeable accounts an inferior flavor to that of Gros Michel, and by now it is known to be to be susceptible to another strain of the wilt Fusarium.

But before UFC and Standard Fruit gave up on Gros Michel and moved to the Cavendish, they –UFC in particular – accumulated huge tracks of lands in Central American countries which they often obtained by shady deals with corrupt politicians. But their land was quickly getting infected with the fungus, and so most of it was soon unsuitable for growing bananas. Nevertheless, the companies refused to give the land back to local farmers to grow other crops on it, hoping that soon they were going to find a cure to the Panama disease and they would then be able to grow bananas again in these fields. This caused a lot of resentment among the local population. Such resentment was generally easily contained, as most Central American countries did not have democratic governments, and UFC made sure that the people in charge were sympathetic to its cause.

However, Guatemala in the 1950s was different. It had a democratically elected government, headed by President Jacob Arbenz, and in 1952 it passed a law that decreed the confiscation of large, unused agricultural plots of land and their redistribution to landless peasants. At the time, UFC controlled 75% of the banana business in Guatemala and its land possessions, four million acres, constituted 70% of all arable land in the country. UFC had no intention of giving up the 40% of their land that the agrarian reform law called for. Nor did they intend to accede to the further demands of the Guatemalan government that they pay the correct amount of taxes that they owed (UFC was turning in fraudulent income and property value numbers to lower the amount of taxes it paid).

What UFC did instead was to obtain help from two other Banana Republics in the neighborhood – Honduras, on the eastern border of Guatemala, and the United States of America further north. UFC was well connected to the political echelon in the United States. The US Secretary of State, John Foster Dulles, had worked for a law firm that represented UFC. His brother, CIA Director Allen Dulles, had been a member of the board of directors of UFC. And Ed Whitman, the main public relationship director of UFC, was married to the personal secretary of President Eisenhower. With these assets, UFC had no problem convincing the American administration (and the American public) that Arbenz was a communist menace to American business interests and must be gotten rid of. In the summer of 1953, Eisenhower authorized the CIA to begin an operation to remove Arbenz from office. This was accomplished by imposing an American naval embargo on Guatemala, and setting up inside Honduras a small, bogus "liberation army" of 480 people, funded, equipped, and trained by the CIA. This "army" was led by a renegade Guatemalan military officer by the name of Carlos Castillos Armas.

The mercenary soldiers under the control of Castillos invaded Guatemala in June 1954, were badly beaten by the Guatemalan armed forces, and retreated back to Honduras. But an American radio station set up by the CIA in Miami kept broadcasting fraudulent reports of ongoing battles and claiming that the Guatemalan army was being defeated. Rumors, generated in part by the Miami broadcasts, began to circulate in Guatemala that the Americans were about to launch an invasion with their own soldiers. The Guatemalan army as well as Arbenz, unable to obtain reliable independent information (UFC controlled the telecommunication systems in Guatemala), began to believe that they were losing the war and that the rebels were advancing on the capital. To save his life, Arbenz resigned and fled to Mexico, and the US (and UFC) were able to install Carlos Castillos as president of Guatemala. Once in office, Castillos[26] quickly rescinded the land redistribution law.

CONCLUSION

As far as human societies are concerned, countries are made of land, and land is where the plants that humans depend on to obtain their food grow.[27] This is why wars are, by and large, about gaining or defending a territory. In this chapter, however, we saw how territories can be gained by the apparently peaceful means of buying them, indicating that commerce could seemingly be the extension of war by other means.

However, the examples given here also demonstrate that the purchase of land by people from outside the country could have far-reaching consequences on the political and social affairs (the two terms are really almost synonymous) of a nation, consequences that are often detrimental to a large number of local residents who may therefore decide to resist such changes by force. For such major consequences to occur, the land purchase must constitute a significant portion of the country, and it has to be used in such a way as to make the land unavailable to the local residents – and the only way to do so is to grow crops on the land, usually with one cash crop dominating.

To date, Zionism in Palestine is perhaps the only historical example in which, by simply buying land without initial military backup in a country in which they were a negligible minority, one ethnic group was able to begin a process that eventually led to the establishment of a state where it now constitutes the dominant group and has complete political control of the country. But the establishment of the state of Israel certainly shows that buying land, establishing homes and settlements, and growing oranges and other crops on it, while seemingly peaceful activities, could lead to major violent conflicts because, regardless of technical and legal aspects of how the land is obtained, possession of land and agricultural activity are fundamental to controlling the country. Thus, land purchasing is fundamentally different from any other kind of commerce, and its political nature becomes overt sooner or later. That land purchasing by Zionists had a political rather than a commercial goal was obvious from the start, not least because they were paying very high prices for the land and their settlements generally were not commercially profitable. Certainly it was clear to many of the local Arab residents, who were mostly farmers and in some cases directly lost their lands and homes, that these activities were going to drastically change the character of their country. They have both politically and violently resisted this change ever since, albeit mostly unsuccessfully.

The acquisition of most of the arable land in Hawaii and its conversion to sugarcane plantations by recent immigrants from the United States of America, who kept their allegiance to their home country (notwithstanding the fact that many were married to local noble women), was the direct reason that Hawaii lost its independence and was annexed to the United States. While this process did not involve a major war, it did not happen completely peacefully either. In the face of mostly peaceful resistance from the majority of the native population, who by then were a minority in their own country because of their susceptibility to lethal introduced diseases and the massive importation of foreign workers to the sugarcane plantations, it was nevertheless the deployment of the US Marines that assured the success of the 1893 revolution by the American expatriates. And it was the continuing implied (and on occasions, applied) force that guaranteed these revolutionaries their control of the country until it was eventually annexed by a militarily even stronger country, the United States – which was their political goal in the first place.

The acquisition of a major part of a country by a political movement – as Zionism in Palestine – and by individuals who only subsequently formed a common cause – as the haole in Hawaii – are cases that led to far-reaching changes in the actual identity of the countries where they occurred. Land acquisition by foreign companies, for the sole purpose of making a profit – as with UFC in Latin America or the two huge rubber plantations that the Firestone company has in Liberia – may appear more politically innocuous, but history shows that they too have led to large amount of political instability and therefore contributed to violent conflicts.[28] Similar activities by companies today, including Chinese logging companies buying forest lands in Papua New Guinea and Saudi Arabian companies buying land in South America, Africa, and even in the United States, for the purpose of growing food and fodder that are shipped back to arid Saudi Arabia, may also appear politically inconsequential for now, but history would argue otherwise.

NOTES

1. An alternative to possessing agricultural land is to have access to the sea for fishing, but large human populations cannot persist and thrive on sea food alone.
2. Lands could be given as a present or dowry in royal weddings, and often the monarchs were not closely related to the people living in these lands. Monarchs could sell land to others, as Napoleon and Tsar Alexander II did when they sold the Louisiana and Alaskan Territories, respectively, to the United States. However, the territories being sold were colonies and not considered part of the homeland. It is often claimed that Mexico sold the land that is now most of the western United States to the US Government, but this was a formality insisted on by the United States after it defeated Mexico in a war, to make the transfer of territory appear legal.
3. Barnes, P. 1999. *A Concise History of the Hawaiian Islands.* Petroglyph Press.
4. Pineapple became an important crop only in the early 20th century, after Hawaii became a US territory. Today, there is very little sugarcane or pineapple farming in Hawaii because of the high cost of land and labor.
5. Dreyfus, an officer of Jewish extraction in the French Army, was falsely accused of spying for Germany and found guilty in a military trial. Many contemporaries felt, and historical evidence later confirmed, that the accusations against Dreyfus were motivated by antisemitism, as was the conduct of the judges.
6. See Grossman, D. 2011. *Rural Arab Demography and Early Jewish Settlement in Palestine.* Transaction Publishers; and Morris, B. 2001. *Righteous Victims: History of the Zionist–Arab Conflict, 1881–2001.* Vintage Books.
7. Most of them were highly religious and spent their days perusing Jewish religious texts, relying for their livelihood on donations from Jewish communities abroad. A small minority engaged in local commerce.
8. At one time or another during history, a substantial area east of the Jordan River ("Transjordan") was considered part of Palestine. However, Transjordan is mostly a desert with few people inhabiting it, and it was not included in the final British Mandate area, the last political entity called "Palestine" before the partition plan was approved by the UN in 1947. It is worth noting too that during Turkish rule, Palestine was part of the Syrian Province in the Ottoman Empire.

9. The First Zionist Congress was held in Basel in 1897 at the instigation of Theodor Herzl to coordinate activities among various Zionist organizations.
10. The first major donor for Jewish settlements in Palestine, Baron de Rothschild, tried to set up a business model for his villages that would allow them to become financially profitable and so not dependent on further donation, but this business model mostly failed. The demand for land on the part of Jewish buyers that was not linked to economic consideration, however, led to an increase in the price that the Arab sellers demanded for it. In fact, it has been shown that the major impediment to land purchasing by Jews was lack of funds, not willing sellers.
11. During Ottoman rule, there were various limitations on the purchase of land in Palestine by Jews and/or non-Ottoman citizens, but they were generally easy to circumvent.
12. Bayer, R.J., et al. 2009. A molecular phylogeney of the orange subfamily (Rutaceae: Aurantioideae) using nine cpDNA sequences. *American Journal of Botany* 96: 668–685.
13. Rouseff, R.L., Perez-Cacho, P. R., and Jabalpurwala, F. 2009. Historical review of citrus flavor research during the last 100 years. *Journal of Agricultural Food Chemistry* 57: 8115–8124; and Tietel, A., et al. 2011. Taste and aroma of fresh and stored mandarins. *Journal of the Science Food and Agriculture* 91: 14–23.
14. Rouseff, R.L. and Perez-Cacho, P. R. 2008. Fresh squeezed orange juice odor: a review. *Critical Reviews in Food Science and Nutrition* 48: 681–695.
15. This was mostly due to the lack of mold spores on the surface of the fruit, the result of the relatively dry climate in Palestine.
16. Porath, Y. 1974. *The Emergence of the Palestinian–Arab National Movement, 1918–1929.* Frank Cass and Company.
17. Jarret, R. L., et al. 1992. RFLP-based phylogeny of the Musa species in Papua New Guinea. Theoretical and Applied Genetics 84: 579–584.
18. A good history of bananas can be found in Koeppel, D. 2009. *Banana: The Fate of the Fruit that Changed the World.* Plume (Penguin Group).
19. Bugaud C. and Alter, P. 2016. Volatile and non-volatile compounds as odor and aroma predictors in dessert banana (Musa spp.). *Postharvest Biology and Technology* 112: 14–23; and Pino J.A. and Febles, Y. 2013. Odour-active compounds in banana fruit cv. Giant Cavendish. *Food Chemistry* 141: 795–801.
20. Goff and Klee (Goff S.A. and Klee, H.J. 2006. Plant volatile compounds: sensory cues for health and nutritional values? *Science* 311: 815–819) have hypothesized that volatiles which are made in the fruit via the degradation of compounds that are important nutrients to animals, such as amino acids and fatty acids, serve as long-distance advertisements to the animals, indicating that such nutrients are present in the fruit. Humans find many esters derived from fatty acids and amino acids particularly appealing (esters are a class of molecules made by combining an alcohol moiety with an acid moiety).
21. Kanazawa K. and Sakakibara, H. 2000. High content of dopamine, a strong oxidant, in Cavendish banana. *Journal of Agricultural Food Chemistry* 48: 844–848.
22. Chagnon, N.A. 2013. *Noble Savages: My Life Among Two Dangerous Tribes – The Yanomamö and the Anthropologists.* Simon and Schuster.
23. Of course, today some apples are exported to far-away countries, too.
24. By extension, the term "Banana Republic" is used to describe any country in which government corruption caused by business interference to enhance their interest to the detriment of the interest of the general citizenry is rife. The term was coined by writer O. Henry in his 1904 book *Cabbages and Kings.* The book was written a few years after his short stay in Honduras.
25. The Colombian Banana Massacre features prominently in the novel *One Hundred Years of Solitude*, written by Gabriel García Márquez, who eventually won the Nobel Prize in Literature for this and other works.

26. Castillos himself was assassinated three years later in another coup.
27. Again, with rare exceptions such as Inuit culture – people who live on land but are dependent for sustenance on animals living in the sea.
28. The Firestone plantation, started in 1926 and occupying a total of 650 square kilometers, 0.6% of the entire territory of the country, has had a major impact on economic and political conditions in Liberia, partially through a loan that Firestone forced the Liberian government to take and repay under very onerous terms (Church, R. J. H. 1969. The Firestone rubber plantations in Liberia. *Geography* 54: 430–437). More importantly, Firestone financially helped Charles Taylor, the convicted war criminal, in his rebellion against the elected government of Liberia and sheltered his forces within the plantation area, in return for being allowed to continue their operation in the country. The civil war that resulted from Taylor's rebellion cost the lives of 300,000 Liberians, and many more were wounded, maimed, and raped. Firestone, originally an American company, was at the time already owned by the Japanese company Bridgestone.
29. Modified from Grossman, D. 2011. *Rural Arab Demography and Early Jewish Settlement in Palestine*. Transaction Publishers.

8 Black Plant Power – Coal and Oil

COAL AND OIL ARE BIOCHEMICALS THAT STORE SUNLIGHT ENERGY

In previous chapters we discussed the myriad types of extant plants that have exerted a huge impact on human affairs, focusing on their contribution to violence. But as described in this final chapter, plants, and other photosynthetic organisms, that have been dead for hundreds of millions of years have also played a crucial role both in the emergence of highly sophisticated and complex human societies in the last few hundreds of years and in the development of mechanized and highly lethal systems and instruments of war. They accomplished this feat by giving rise to coal and petroleum.

As we have seen in previous chapters, the human body is a machine that requires the input of organic compounds to grow and maintain its structure, and to use their energy to power metabolic processes and achieve movement via the action of muscles. Other living machines that humans use – beasts of burden – also require organic material input for the same purposes.

For most of history, humans performed work (the general definition of work in physics is the application of force to move a physical object) with their own muscles or through the use of beasts of burden. To operate, muscles obtain energy by metabolizing organic compounds that the animals ingest. Throughout history, humans also kept inventing non-living machines. Many of these machines were simply powered by human muscles or the muscles of other animals, but some were powered by energy contained in wind and water. With the developments of the steam engine in the 17th century and the internal combustion engine in the 18th century, a large amount of work achieved by humans began to be accomplished by inanimate machines powered by burning organic materials. The materials being used were mostly coal and petroleum ("rock-oil," also known as crude oil) and to a lesser extent dried wood obtained from cutting down live trees.

Coal and petroleum, both known as "fossil fuels" for reasons described below, consist mostly of carbon atoms that are in a chemically unoxidized state, namely that they are not linked to other atoms such as oxygen whose nuclei exert a stronger attraction force on the carbon atoms' electrons. When these carbon atoms in coal and oil, which are usually bound to hydrogen atoms, come in contact with oxygen molecules, they "burn" – some of their electrons become tightly bound to the oxygen molecules, although these electrons remain connected to the carbon nucleus as well. This chemical reaction, which links the carbon and oxygen atoms, usually results in the formation of the compound carbon dioxide (CO_2), in which one carbon atom is linked to two oxygen atoms, and water (H_2O), in which the hydrogen atoms that the carbon atom released when it combined with oxygen themselves bind to another oxygen atom (by sharing their own electron with the electron-loving oxygen nucleus).

While the burning reaction, chemically called "oxidation," requires an initial input of energy to get started, once the reaction gets going a large amount of energy is released, leading to an increase in temperature of the reactants and products, as well as the surrounding material. This released energy can be used to boil water to produce steam, the gas form of water. In the steam engine, the expansion of steam as it is produced and is further heated in a confined container exerts the force to move the pistons of the engine or the blades of the turbine. In the internal combustion engine, the expansion of CO_2 and H_2O vapors produced from the combustion of the liquid fuel in a closed container pushes the pistons directly. In a jet engine (a version of the internal combustion engine) mounted on an airplane, the CO_2 and H_2O gas are released from the rear of the engine, pushing against the resistance of the ambient air and thus propelling the plane forward.

A steam engine can be operated by burning wood, and an internal combustion engine could be powered by burning oil or alcohol produced from crop plants. But from the beginning of the industrial era, coal, and later petroleum, were greatly preferred, as they constitute dense, concentrated forms of fuel that are found in seemingly unlimited amounts. Furthermore, such energy resources are present underground and therefore do not compete with agricultural use of arable land for the production of food and, at any rate, are much cheaper to obtain than biologically based fuels.

Coal has also assumed an essential role in the manufacturing of iron and steel, a process called smelting. Developed thousands of years ago, smelting begins with iron ore, which consists of oxidized iron atoms bound to oxygen atoms (basically, iron rust). To extract the iron, it is necessary to break the bonds between metal and oxygen, and this requires a source of electrons. This process also requires high temperature. Combustible carbon serves both roles admirably – some of the carbon atoms can combine directly with atmospheric oxygen (i.e., burn) and produce carbon monoxide (CO), CO_2, and heat, and some of the CO molecules (the carbon in CO still has two available electrons unbound to oxygen) that are formed can combine with the oxygen atoms that are currently bound to the iron atoms, releasing the iron atoms from the grip of the oxygen atoms and also producing CO_2 in the process. To achieve the extremely high temperatures needed, the smelting process takes place in a chamber, called a kiln, where the concentrated carbon source is burned under conditions of high oxygen levels. Originally, the source of carbon used was not coal but charcoal, which is prepared by partially burning wood (by limiting the amount of oxygen) so that what remains are mostly concentrated, solid carbon atoms, like in coal. But as firewood for charcoal production became scarce and expensive, coke, a specially produced powder of coal, took the place of charcoal in iron and steel production.

The net result of the smelting process is that the iron atoms, now in sole and complete possession of all their electrons, are no longer in the rusted (i.e., oxidized) state, and they recover their metallic properties and appearance. Because of the high temperature of the kiln, the newly formed iron is in a liquid state, and it flows out of the kiln and then solidifies as it cools. The initial metal obtained this way, called pig iron, contains up to 5% carbon, and is neither very strong nor flexible. Additional heating in the kiln under the proper conditions (the right temperature and the use of special coke that has low levels of impurities) leads to the production of steel, which has up to 2% carbon, or wrought iron, which has none.

FORMATION OF COAL

Coal deposits are found all over the world (Table 8.1). Most of them were formed during the aptly name Carboniferous geological period, 360–290 million years ago, when the climate on Earth was much warmer.[1] The land mass at that time was covered with dense forests of plants dominated by fern trees and newly evolved conifer trees that busily harvested light energy and used it to make organic molecules. As we saw in Chapter 6, much of the mass of the plant body is made of two polymers, cellulose and lignin, that are linked together, and by the Carboniferous few, if any, microbes had evolved the ability to degrade these entwined polymers. Therefore, much of the dead plant material – representing millions of years of harvested sunlight energy – accumulated on the ground and in shallow ponds, eventually to be covered by sediment. The underground burial excluded oxygen and further preserved the plant material.

As additional sediment accumulated on top, the pressure and temperature to which this organic material was subjected rose. Under these conditions, several things occurred. First, water remaining in the dead plant material (all living cells are mostly water by mass) was pushed out, drying the material. Second, the high temperature caused a chemical reaction, called dehydration, in which oxygen and hydrogen atoms bound to the carbon skeleton of the organic compounds were detached from

TABLE 8.1
Annual Coal Production of the Top Ten Producing Countries, in 2015[a]

	Country	Production (in million metric tons)[b]	Continent
1	China	2610	Asia
2	United States	650	North America
3	India	405	Asia
4	Australia	392	Australia
5	Indonesia	344	Asia
6	Russian Federation	262	Asia/Europe
7	South Africa	204	Africa
8	Columbia	80	South America
9	Poland	77	Europe
10	Kazakhstan	68	Asia/Europe

[a] Data was obtained from British Petroleum: Statistical Review of World Energy, 2016. www.bp.com/content/dam/bp/pdf/energy-economics/statistical-review-2016/bp-statistical-review-of-world-energy-2016-full-report.pdf

[b] Amounts in actual tonnage of coal were obtained by converting the amounts given in the original table in "oil equivalents"; 1 metric ton of "oil equivalent" is roughly 1.43 metric ton of coal.

the carbons and combined with each other to form water (H_2O), and this water was removed too from the organic material by the heat and pressure. Finally, molecules with long carbon chains broke down ("cracked") into smaller molecules, sometimes as small as containing just one carbon atom (forming methane gas, CH_4), or they simply rearranged into a regular array in which each carbon atom was at an equal distance from several other carbon atoms, forming a solid crystal.

The end result of this process, which takes many millions of years to complete, is almost pure, crystalline carbon material called graphite. However, very few of the coal deposits around the world have reached this stage. The youngest form, and the one with the lowest concentration of carbon, is called peat, and the next stage in the development of coal is lignite, containing a higher concentration of carbon (60%–70%). Sub-bituminous and bituminous coals, the most common deposits of coal and the types used most often for various industrial purposes, represent the next two stages in coal development. Anthracite, with a 91%–98% carbon concentration, represents a stage of coal development that is just short of graphite, and, like graphite, is also relatively rare. Anthracite forms a hard, black rock that is difficult to ignite, but, once it catches fire, burns with little smoke. Since coal is made from plant material, it naturally also contains appreciable amounts of sulfur and nitrogen, although their concentrations typically decrease as the coal matures.

FORMATION OF PETROLEUM

Petroleum, often called crude oil, was formed by analogous processes to those responsible for the formation of coal, except that the starting organic material was marine organisms. Just like life on land, life in the ocean is also dependent on photosynthetic organisms that harvest light energy and CO_2 to make organic material, and just like on land, marine photosynthetic organisms constitute the wide bottom of the food pyramid in the ocean and the largest proportion of biological mass, by far. However, most of the photosynthetic organisms present in the oceans several hundred years ago, or even today, are not actually classified as plants. Some were (and are) photosynthetic bacteria and green and red algae – botanically speaking not plants but sharing ancestors and evolutionary history with plants. By mass, the majority of the denizens of the oceans for the last couple of hundred million years have belonged to a group of microscopic life forms called diatoms.

Diatoms are single-celled organisms that are not closely related to plants, based on comparisons of their DNA sequences with those of plants and non-photosynthetic organisms. But diatoms have one characteristic that makes them similar to plants – they photosynthesize. Their ability to carry out photosynthesis is due to having a structure similar to a plant chloroplast inside the diatom cell. It is believed an ancestor of the present-day diatoms swallowed a single-cell chloroplast-containing algal cell about 200 million years ago and the ingested chloroplast, itself the remnant of a photosynthetic bacterium that the ancestor of the alga had ingested, continued to live and function inside the original diatom cell. Furthermore, the progeny of the diatom cell that swallowed the algal chloroplast have continued to have live-in, functional chloroplasts to this day because diatom cell division (the method of diatom reproduction) became synchronized with chloroplast division (the method of chloroplast reproduction), so that after each diatom cell division each of the two daughter diatom cells receives a chloroplast.[1]

Petroleum "fields" – underground areas of porous rock saturated with oil and gas at various depths, but mostly less than 5,000 meters below the surface – are also spread around the world, albeit unevenly. There are major oil and gas fields in North and South America, Asia, Europe, and Africa, although 65% of all proven oil reserves are concentrated in the Persian Gulf, the site of a branch of the ancient Tethys Sea. Most (70%) of the oil that we now pump out of the ground was formed during the Mesozoic age (252–66 million years ago), with about 10% forming earlier and the rest since.

Average earth surface temperatures during most of the past were generally higher than today. With the breakup of the supercontinent Pangaea during the Mesozoic age, shallow seas such as Tethys began to form. These seas contained a multitude of thriving photosynthetic organisms, such as algae and diatoms, that lived in the upper water layer and captured light energy and used it to make organic compounds from CO_2 and H_2O. There was very little upwelling in these seas, so the water at the bottom had practically no free oxygen (which is produced during photosynthesis when the hydrogen atoms in H_2O are separated from the oxygen), and when living organisms died and sank to the bottom, the organic debris could not be decomposed there by microorganisms (oxygen is required for the breakdown process). Eventually, the dead organisms were covered by more and more inorganic sediment, often brought to the seas by rivers and the rest by wind and shallow currents. As with coal formation, this buried marine organic material began to experience heat and pressure and eventually transmogrified into oil.

One important difference between the dead land plants that gave rise to coal and the dead marine organisms that gave rise to oil was that the latter were almost completely lacking in lignin and cellulose, the bulk of the cell walls of land plants. Instead, the marine detritus consisted of a much higher proportion of cell membranes, which mainly consist of lipids, than that found in the mass of dead land plants. Lipid molecules, which are long carbon chains bound to mostly hydrogen atoms, contain much fewer oxygen atoms compared to cellulose and lignin (see Chapter 2 for the chemical difference in oxygen content between lipids and sugars). Without oxygen, the dehydration process described above for coal, which removes oxygen and hydrogen atoms bound to an organic molecule in an obligatorily 1:2 ratio, has played a much smaller role in petroleum formation. Thus, as the pressure and temperature mounted on this organic material, free water from the dead cells was eliminated as in the case of coal formation, and carbon chains were cracked and shortened, again as in the case of coal, but the final product still had more hydrogen bound to the carbon than in the case of coal.

Such hydrogen-rich carbon chains, called hydrocarbons, are what crude oil mostly consists of (as is the case for coal, petroleum also contains low but various levels of sulfur and other elements found in living cells). The majority of the hydrocarbon molecules in mature petroleum are typically liquid under the temperature and pressure conditions prevailing in the oil field, and the low percentage of heavier hydrocarbon molecules present, whose physical properties would make them solid in pure form under these conditions, are actually dissolved in this crude oil liquid. Overall, typical petroleum liquid is still less dense than the water that was removed from the organic matter when the oil was formed, so crude oil is generally found floating on top of a water layer. The very short hydrocarbon molecules, such as methane, form a separate gas phase above the hydrocarbon liquid.

THE CRUCIAL ROLES THAT COAL AND OIL HAVE PLAYED IN RECENT HUMAN HISTORY

Coal was used as a source of heat by the ancient Chinese and by the Romans in England, but its real influence on historical events, both peaceful and violent, began with its role in the Industrial Revolution. In fact, it can be convincingly argued that coal was a main cause of this revolution. More specifically, the need to obtain and ship increasing amounts of coal was the driving force behind the development of the coal-powered steam engine as well as the railway transportation system, and these, once developed, were used for other industrial processes.

Why the Industrial Revolution occurred when and where it did is an extremely complicated story that has been extensively researched,[2,3,4,5] and only a brief description of the role of coal in it will be attempted here. In England, once the Roman Empire collapsed, the use of coal for heating ceased and this knowledge fell into obscurity and was rediscovered only in the Middle Ages. As the human population on the British Isles grew, the natural forests that were used to obtain wood for heating and cooking were diminishing, and therefore the price of wood (and of wood-derived charcoal, used in large amounts for smelting) was going up. Coal layers near the surface and showing up on hillsides, particularly near Newcastle on the east coast north of London, were easily accessible, and such coal began to be mined by locals. Initially the fuel for only poor people, as firewood became scarce and therefore even more expensive, more people switched to using cheap coal. Soon most of the houses in London, which could receive coal shipments by boats from Newcastle, were heated with coal.

As the easily accessible coal seams were exhausted, technical problems arose. One of them had to do with flooding in the mines. The coal layers often extended farther down below the water table, and as the coal was mined the empty space filled with water, drowning miners and impeding access to the remaining coal. Various methods were tried to remove the water, some relying on a conveyer belt of buckets installed in shafts and powered by animals. In the early 18th century, the Englishman Thomas Newcomen developed and installed the first true steam engine that ran a vacuum pump to remove water from a coal mine. The steam was conveniently generated by burning coal.

So the coal-fired steam engine was developed to mine more coal. The steam engine itself was made from iron and steel, which in the 18th century were manufactured in a process still using mostly charcoal. But soon the increased demand for these metals outstripped the availability of firewood, and coke was developed from coal to replace charcoal in the smelting process. Toward the end of that century, the steam engine was greatly improved by another Englishman, James Watt, and steam soon replaced water as the motive force to run factory machinery, particularly in the textile industry. Water power was always quite limited in England, both in capacity and location. The availability of steam engines to run industrial machinery led to the fast and furious industrialization of England and many other countries in Europe, the United States, and beyond. Finally, in the early 19th century, English engineers developed steam locomotives that ran on steel rails and pulled cars full of – what else – the bulky coal from the mines to the cities and factories where the coal was needed. The iron steamship was soon developed too.

To summarize the role of coal in recent history, during the 19th century coal was the main source of energy for factory machinery and for rail and sea transportation systems, and even for illumination (with coal-derived gas). These roles have been, for the most part, supplanted by petroleum by the first half of the 20th century. However, coal still plays a substantial part around the world in the generation of electric power, which is an increasingly more common form of energy, being used to run both electric motors of various devices and electronic systems such as computers, televisions, and many other "smart" machines. Coal has also been essential, up to the present date, for the production of steel, much of it going to the direct construction of weapons and otherwise an essential element of both peaceful and military industries and infrastructure.

The petroleum industry began its modern development in the middle of the 19th century with a focus on producing kerosene for illumination, replacing both coal and sperm whale oil. However, with the widespread adoption of the internal combustion engine, which uses one fraction of petroleum or another as the energy source,[6] at the beginning of the 20th century, the paramount role of crude oil and natural gas was to serve as the major motive force in industrialized countries, replacing coal. Various crude oil distillation products are also used as chemical feedstocks to manufacture a large number of industrial products, from synthetic rubber (Chapter 6) to plastics, pesticides, pharmaceuticals, asphalt, etc.[7]

With coal and petroleum having been, by far, the most common energy sources for human industrial activity in the last 200 years, and coal being a major ingredient in the production of steel, it goes without saying that coal and oil have played major roles in many aspects of war during this period, from armaments to transportation of troops to the motive force of mechanized military vehicles and devices. And the essentiality of these two materials certainly led to many military campaigns (in addition to peaceful diplomacy, or instead of it) to secure geographical locations where these materials can be obtained. Since it would take too much space to recount the entire military history of recent times, only a few salient and significant events will be illustrated here.

Given the role of coal in the birth of the Industrial Revolution in Europe, it is not surprising that coal was an important factor in the political development of the continent, and in particular Germany. In the west of Germany, two important coal areas, the Ruhr and the Saar (see Figure 8.1), were instrumental for the industrial development of Germany, both before and after its unification in 1871. After World War I, the Ruhr area was declared a demilitarized zone by the victorious allies in the 1919 Peace Treaty of Versailles, and in the early 1920s French and Belgian troops occasionally invaded the area and interfered with coal and steel production in order to pressure Germany to pay the steep war reparations they were owed. The peace treaty also stipulated that the Saar area to the south of the Ruhr would be separated from

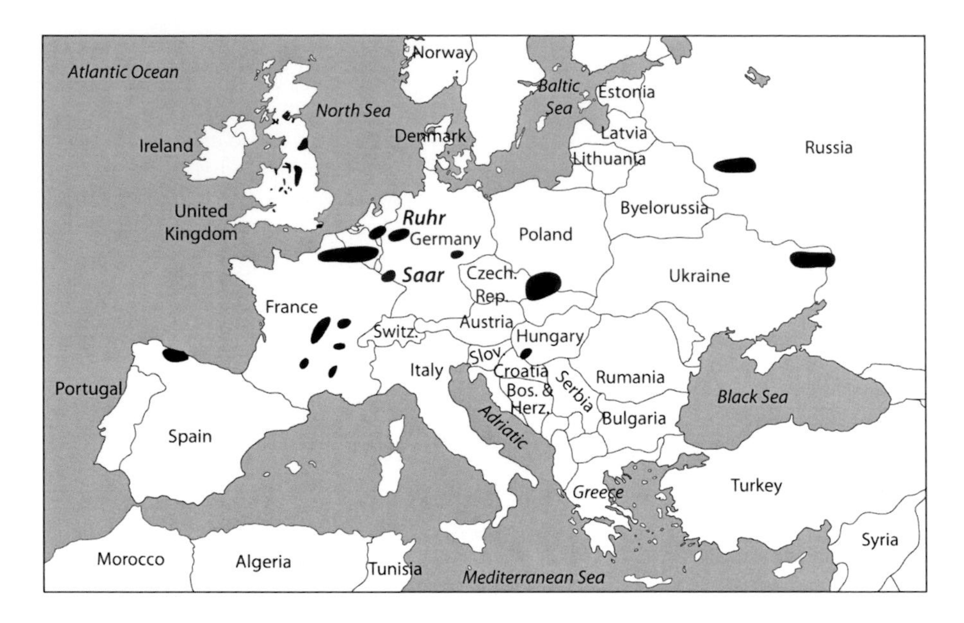

FIGURE 8.1 Locations of major coal fields in Europe (dark areas), and the locations of the Saar and Ruhr areas of Germany.

Germany and be administered, as the Territory of the Saar Basin, jointly by England and France. The loss of control of both the Ruhr and the Saar, imposed by the western allies to prevent Germany from rearming, put a real crimp on Germany's ability to recover economically from World War I and increased the resentment among the German population against the allies.

In 1935, the Territory of the Saar Basin held a plebiscite and the population overwhelming voted to reunite with Germany, a decision that was promptly implemented. In addition, in 1936, Hitler ordered his troops back into the Ruhr in violation of the peace agreement. The failure of the Western countries to force the German army to retreat basically gave the green light to Hitler to launch his rearmament plan in preparation for the German military conquest of Europe. The coal and steel industries in the Ruhr and the Saar played a major role in this plan. When petroleum supplies in Germany during the war became limited, the Germans also used coal as a source to make synthetic fuel and other products normally derived from petroleum, such as synthetic rubber.

During World War II, the Ruhr and the Saar were bombed heavily by the Allies in order to impede coal, steel and synthetic oil production. After the war, France attempted to keep control of both the Ruhr and Saar areas.[8] The allies allowed France to keep the Saar as a protectorate for a while. But in 1955, a plebiscite to make Saarland an independent country was rejected by the local population, and a subsequent agreement negotiated between France and Germany allowed the Saar to join the Federal Republic of Germany (West Germany) in January 1957. In the Ruhr, France was not able to obtain agreement from its allies to maintain long-term, direct physical control after the war as it did in the Saar. In response, in 1950 the French statesmen Jean Monnet and Robert Schuman devised a creative solution to keep the

Germans from regaining control over the Ruhr (and Saar) coal fields. Their solution called for the creation of an entity to control the production and sale of coal and steel throughout Europe whose authority will supersede national sovereignty.[9] This entity would be managed jointly by the countries that joined it, with each country representative having a veto power over any decision. Initially, six states – France, Germany, Italy, Belgium, the Netherlands, and Luxemburg[10] – agreed to establish this organization, named the European Coal and Steel Community (ECSC), which eventually evolved first into the European Economic Community (EEC) and later into the European Union (EU).

On the other side of the globe, coal played a major part in the military colonial expansion of Imperial Japan in the late 19th century and the first half of the 20th century. The Japanese Islands are poor in fossil fuel resources. As Japan began industrializing in the second half of the 19th century, it felt a keen need to obtain coal, among other natural resources. Certainly, the coal deposits of northern Korea were a major factor in Japan's military expansion into Korea, which caused the latter to become (against its will) a protectorate of Japan, formalized in the Japan-Korea Treaty of 1905. North and west of Korea lay Inner Manchuria,[11] a region even richer in coal than Korea. Japan's military forces invaded Inner Manchuria in 1931, establishing there a client state they called Manchukuo and installing the deposed Chinese Emperor Puyi as the titular "Emperor of Manchukuo" to run it under their tight control. Japan then proceeded to set up a coal and steel industrial complex in Inner Manchuria, and it used the territory and industrial resources developed there (including a rail system) as a base for invading the rest of China in the late 1930s.

Much as coal was, for a brief period in history, the most important energy resource for preparing and mobilizing land and sea military campaigns, its supremacy was soon eclipsed by petroleum. The year this happened can clearly be dated as 1913, when, at the instigation of First Lord of the Admiralty Winston Churchill, the British Navy converted from coal- to oil-powered ships. While Britain had large reserves of coal but no domestic sources of petroleum, the much greater efficiency of petroleum clearly justified the risk that unsecure access to petroleum posed. Since 1913, it is likely that no war has been fought around the world – perhaps with the exception of some tribal skirmishes in the Amazon – that was not highly dependent on petroleum products. All military transportation devices for moving troops on land, sea, and air are powered by petroleum products, as are mechanized military weapons such as tanks, ships, jet planes,[12] rockets, and missiles. Furthermore, the most widely used military explosive, TNT (2,4,6-trinitrotoulene), is made by modifying toluene, a distillation product of petroleum.[13]

The move to depend on non-domestic petroleum sources for civilian industrial development and military campaigns was risky for many oil-poor developed countries. While the United States and Russia had, at the time (and Russia still does today), enough domestic oil supplies, other European military powers, as well as Japan, did not, and this fact could not be ignored by their statesmen and military strategists. Naturally, these countries acted to secure their oil supplies, and indeed a case can be made – and such a case has been made in numerous books dealing with this topic – that petroleum became not just a resource that made fighting wars possible, but one that many modern wars were waged specifically to gain control of.

Without going over the entire world history since 1913, only a few major mileposts will be mentioned here. The main goal of England in its campaign in the Middle East during World War I was to ensure control of the oil fields in the Persian Gulf. For this purpose, the state of Iraq, a completely new entity with no historical foundation, was created to cover most of the territory where oil fields were then known to exist. This artificial country, combining three major religious groups (Shi'ites in the south, Sunnis in the center, and Kurds in the north), has had a long and violent internal history ever since. Furthermore, the arbitrary borders drawn by the British and their accomplices the French after World War I have made the entire region unstable to this day, leading to various local wars, many of them over oil. Most of those conflicts flared up after World War II, though, so before we describe these Middle Eastern wars, a short description of the role of petroleum in the intervening years is in order.

An early demonstration of how the desire to control oil could foment violence was the Chaco War, fought between Bolivia and Paraguay from 1932 to 1935, the last major war on the South American continent up to this date. The part of the Chaco region that was in dispute was located at the confluence of the countries of Bolivia (northwest of it), Paraguay (southeast), Brazil (northeast), and Argentina (southwest). This low-lying area was at the time (and still is) sparsely populated, and its ownership contested between Paraguay and Bolivia, both land-locked nations that had lost territories to other neighbors in 19th century wars. The legal details of the territorial dispute, which will not be recounted here, were complex and stretched back many years. What gave the two countries the incentive to aggressively pursue their claims was the belief that the area was rich in petroleum reserves, similar in scope to the oil fields that had already been discovered in the eastern foothills of the Andes inside Bolivia proper.

The three-year war caused as many as 100,000 casualties to both sides combined and ended in victorious Paraguay gaining control of most of the Chaco region. Ironically, though, no oil had been discovered in the area until 2012, and even with the recent findings it is not yet clear how much oil there really is in the Paraguayan Chaco. At any rate, there has been no commercial oil production in Paraguay yet, and the country imports all its oil from neighboring South American countries.

During World War II, Germany, the main aggressor in the European theater, initially had no significant oil deposits under its control, while Britain had control of Middle East oil and Russia had ample domestic supplies. Furthermore, the United States was a net exporter of petroleum, and it could and did help Britain with oil supplies during the war. Germany was very cognizant of its disadvantage in lacking secure oil supplies, and in preparation for the war it had built a major chemical industry to make synthetic liquid fuel from coal (and as mentioned earlier, many of these installations were in the Ruhr and Saar regions), but the process proved extremely expensive and inefficient. Therefore, when Germany embarked on its invasion of the Soviet Union in the summer of 1941, its Army Group South (one of the three main forces, the other two being Army Group Center and Army Group North) aimed to advance all the way to the shores of the Caspian Sea to take control of the Baku oil fields in Azerbaijan, the main Russian oil fields at the time. On the way to Baku, the Germans easily captured and began exploiting the Rumanian oil fields, with the consequence that the Rumanian oil installations soon became a major target for the Allies' bombers. But in November 1942 the German push to the Caspian Sea was decisively and irreversibly halted by Russian forces west of the Chechnian capital of Grozny, a few hundred miles from the Caspian Sea. The Germans had put enormous resources into this campaign, to the detriment of their two other Army Groups, and the catastrophic losses that Army Group South sustained throughout the campaign, most notably the loss of an estimated 800,000 German troops and those of their allies in the Battle of Stalingrad, doomed their entire campaign in Russia and guaranteed their overall defeat in the war.

While in Europe petroleum was essential for the conduct of World War II, in the Pacific petroleum was the immediate cause of the war that Japan declared on the United States on December 8, 1941 (Japan time). As described earlier in the chapter, Japan's industrialization, which began in the second half of the 19th century, was closely linked to, and dependent on, its military expansion into Asia. This expansion brought Japan into conflict with the old local empire, China, as well as with the newer imperialistic powers in the neighborhood – Russia, the United States, England, France, and the Netherlands. After Japan's occupation of Inner Manchuria (previously described), it continued to expand south, launching a major attack on China in 1937. In order to subdue China, Japan imposed a blockade that caused it to come into direct conflict with the Western powers, who desired free trade access to the mainland. Starting in 1938, the United States and other Western countries began to impose trade restrictions on Japan, which was particularly dependent on iron and petroleum purchases from the United States. In response to the trade restrictions, the Japanese government, undeterred from its military campaign in China, signed the Tripartite Pact with Germany and Italy in 1940, which guaranteed mutual assistance in case any party to the agreement was attacked. And to further tighten its blockade of China, the Japanese military next occupied major naval bases and airfields in French Indochina, which they were able to do peacefully because the French Vichy government collaborated with the Germans.

The Japanese take-over of the military bases in Indochina led the Americans to tighten their embargo on petroleum to Japan, with a complete oil embargo announced on August 1, 1941. At this point, Japan's military strategists knew they had only limited time, which they estimated at two to three years, to find alternative petroleum resources before they ran out fuel. The plan they developed was for Japan to expand south and to occupy the oil fields in Brunei (on the island of Borneo), controlled by the British, and Sumatra (part of Indonesia), controlled by the Free Dutch government. Neither the British nor the Dutch were friendly to the Japanese, and in fact the Dutch had refused a request by Japan to sell them large quantities of oil. But for the Japanese military expedition to be successful, the Japanese generals felt they first had to neutralize the Americans, who controlled the Philippines and could easily intercept the Japanese forces or come to the aid of the British and the Dutch. With a knack for long-term planning, the Japanese, rather than simply attack the Americans in the Philippines and see quick American response from their forces elsewhere in the Pacific, decided on a more thorough plan – to knock out the main American navy and air bases in the Pacific Ocean, situated at Pearl Harbor and elsewhere on the island of Oahu in Hawaii.

As is well known, the long-term consequences to Japan of their devastating surprise attack on December 7, 1941 were quite dire. While they went on to overrun most of the Far East and obtain control of the oil fields in Brunei and Sumatra, as well as the rubber plantations in Malaya, they were not able to withstand the overwhelming superiority in military and economic resources and personnel that the United States was able to marshal against them, and within four years they were decisively defeated. Perhaps less appreciated today is the fact that soon after the attack, on December 11, Hitler declared war on the United States, forcing the country to join the war in Europe, a move that the United States was not yet ready to make voluntarily. Hitler's motives for doing so are still being debated by historians. Perhaps he felt he was required to do so because of the Tripartite Pact, although technically this was not the case, since Japan was the aggressor rather than a victim of an attack. Nonetheless, the entrance of the United States into the European theatre of World War II guaranteed the defeat of Germany as well.

World War II made clear the strategic importance of oil. After the war, the two main sources of oil outside the Soviet Union and the United States were located in the Persian Gulf and in Venezuela. In addition, with the enormous peaceful industrial development that the United States was going through after the war, it became clear to American long-term strategic planners that the United States would soon become a net importer of petroleum (which it did in the middle of the 1950s). The onset of the Cold War, pitting the "Free World" and its postwar military alliance, the North Atlantic Treaty Organization (NATO), against the Soviet Union and its allies, accentuated the strategic importance of petroleum. This led to the issuance of American National Security Council Report 138/1 on January 6, 1953, during the last days of the Truman administration. The top-secret report stated in part:

> Since Venezuela and the Middle East are the only sources from which the free world's important requirements for petroleum can be supplied ... it therefore followed that nothing can be allowed to interfere substantially with the availability of oil from these sources to the free world.

This policy was soon put to the test. In mid-1951, the democratically installed government of Iran, led by Prime Minister Mohammad Mosaddegh, nationalized British Petroleum (BP), the oil company that had monopoly over the oil fields in Iran and in which the British government had a majority stake.[14] This was done after BP ignored the Iranian government's demands to increase the payment of royalties to Iran and to improve the working conditions of the Iranian workers in the company. In response, the British government initially imposed an economic embargo on Iran, but in November 1951 Churchill became prime minister again and the British government became more belligerent toward Iran. An attempt by the British to foment a coup against the Iranian government in 1952 was so clumsily carried out that not only did it fail, but the British involvement became public and the Iranian government severed diplomatic ties with the United Kingdom. Nor did the British receive much sympathy from President Truman, who, although he did eventually sign NSC 138/1, still believed in honoring international law.

However, when the new American Administration of President Eisenhower took over in early 1953, things changed drastically. The British and their allies in the American administration, in particular the Dulles brothers, in charge of the State

Department and the CIA (see Chapter 7), were able to convince Eisenhower that Mosaddegh was either leaning toward better relationships with the Soviet Union or at least was not sufficiently loyal to the cause of fighting the spread of communism. Therefore, there was the danger that he would allow the Soviet Union to gain control of Iran's oil fields. This line of argument was so successful that the Americans took it upon themselves to do the job of removing Mosaddegh. The successful, bloody coup, orchestrated by the CIA, occurred in August 1953, in the same year that the CIA also acted to remove the Guatemalan President Arbenz to make that country safe for the United Fruit Company (Chapter 7).

In 1956, the British instigated another military campaign in the Middle East over control of petroleum, this time its transportation. In July 1956, following a long and complicated chain of events, the President of Egypt Gamal Abdel Nasser nationalized the Suez Canal Company, which was a publicly held company with mostly French and British stockholders. At the same time, Nasser prohibited the movement of Israeli ships through the canal and also declared a military blockade in the Gulf of Aqaba to ships bound to the Israeli port of Eilat on the Red Sea. These combined actions isolated Israel from its main oil supplier Iran and made the supplies of oil from the Persian Gulf to Europe uncertain and under the control of a leader who appeared not to be friendly to the West. In response, Britain, France, and Israel banded together and in late October 1956 (almost) simultaneously attacked Egypt.[15] The Israeli forces advanced from the east and captured the entire Sinai desert, quickly reaching within a few miles of the canal, while British and French troops captured the area around the canal. This time, however, the US administration under President Eisenhower condemned the invasion – the Soviet invasion of Hungary was happening at the same time, and the Unites States felt it had to be consistent and condemn all foreign aggression – and the British, French, and Israeli forces withdrew from all Egyptian territories that they had captured. The Suez Canal remained nationalized and inaccessible to Israeli ships, but ship movement through the Red Sea to Eilat was allowed to resume (until Nasser blockaded Eilat again in 1967, precipitating the Six-Day War in June that year).

Back to Iraq: in 1980, Iraq, under the leadership of its dictator Saddam Hussein, invaded Iran with the aim of gaining control of additional oil fields, which it ultimately failed to achieve. In 1990, Hussein's Iraq successfully invaded Kuwait for the same reason. While the United States covertly helped Hussein in his war against Iran, after his invasion of Kuwait in 1990 the United States decided that allowing Hussein to control 20% of the world's proven oil reserves was not in its interest, and it convinced the United Nations to authorize a military campaign to evict Iraqi forces from Kuwait. The United States led this military effort and indeed was able to destroy the Iraqi army and to liberate Kuwait, although Hussein remained in firm control of Iraq. In 2003, the United States launched a second war against Hussein's Iraq with the explicit aim of removing him from power. Although oil was not the official *casus belli* of this war, many political observers felt that American policy makers believed that the continuation of the Hussein regime and its influence in the region, particularly as it pertains to oil exports, was not in its national interest, and that this thinking contributed to the decision to attack.

COAL, COAL MINERS, AND SOCIAL STRIFE

When counted on per energy unit, work in the oil fields always required less human labor compared to coal, although recent automation trends in the coal industry are narrowing the gap somewhat. At any rate, work inside coal mines has always been extremely dangerous, and even today many thousands of coal miners are killed in accidents every year around the world.[16] The inherent danger in underground mining of any material has meant that throughout history, mine owners tended to employ slaves, and outright slave labor in coal mines was not unknown (for example, in Scotland in the 17th and 18th centuries and in Nazi Germany during World War II). Most often, though, individual mine owners and corporations took advantage of poor and uneducated people in ways that were more or less legal at the time. For example, in the United States in the 19th and early 20th centuries, miners had to live in "company towns" built and operated by the mining company. The workers were charged for housing of course, and sometimes even for their working tools, and they had to buy their food and other necessities in stores run by the company, often at inflated prices. In fact, many times the workers were not even paid with real money, but with "scrip," a piece of paper that could be exchanged for goods only in the company's stores. As a result, coal miners were often in legal debt to the mining company that employed them and had little freedom to quit their job. Moreover, if a miner became uppity and requested a higher salary or better working conditions, or worse, if he tried to band together with other workers to make such a request, the company could, and often did, evict him and his family from the abode he was occupying. This was certainly a good method to obtain and keep very cheap labor.

As coal mining expanded in the United States in the second half of the 19th century, with mines in Pennsylvania, the Appalachian states of West Virginia and Tennessee, and further west in Illinois and Colorado, the industry developed additional ways to keep its labor cost low as they fought attempts by the workers to unionize. In Tennessee, for example, mine owners began to employ convicts, whom they "leased" from the state. Everywhere, mine owners hired private guards who were often organized as well-armed militias and whose job was to intimidate and evict troublesome miners. Such intimidation sometimes extended into outright murder. The miners often retaliated, and between roughly 1890 and 1920 there were multiple lethal skirmishes between miners and the guards employed by owners, with occasional small "massacres" occurring in various states, for example the "Ludlow Massacre" in Colorado in 1914.[17] This streak of violence reached it apotheosis in the Battle of Blair Mountain fought in late August and early September 1921 in Logan County, West Virginia.

The Battle of Blair Mountain occurred when about 10,000 mine workers, members of the United Mines Workers (UMW) union, decided to march to Mingo County in West Virginia to confront the mine companies operating there.[18] To get there, they had to march through Logan County. The local sheriff, Don Chafin, an anti-unionist, decided to physically block their way. He assembled a force of 3,000 men, consisting of state police, deputies, and citizen militia equipped with machine guns and rifles, and placed them on Blair Mountain, a promontory overlooking the route that the miners had to take on the way to Mingo County. The miners themselves, many of them veterans of World War I, came well-armed with various guns. As the miners approached, the men on Blair Mountain opened machine-gun fire on them, and the miners fired back and tried to storm the promontory. The most intense stage of the battle went on for two days, August 31–September 1, and involved three planes that dropped ineffectual bombs and tear gas canisters on the miners. While about 100 people died in these skirmishes (two-thirds of them miners), the outcome was a draw with neither side budging. The next day, September 2, army troops dispatched by President Harding arrived to put an end to the fighting, and they were successful in quieting things down within a few days. The troops also arrested almost a thousand miners, many of whom were later sentenced to jail time. None of the combatants on the other side were arrested.

OIL AND SOCIAL STRIFE

The oil industry provides somewhat different opportunities for social conflict than coal does. The established oil industry has always required a small but highly trained work force for oil exploration, drilling, and production. Since oil fields are often located in relatively underdeveloped countries, the development of oil resources has mostly been undertaken by multinational companies whose headquarters are in one of the more developed countries. These companies often obtain "concessions" to exploit oil fields in underdeveloped countries, with the profits accruing mostly to the shareholders of the companies. The local inhabitants benefit very little from such activity, and they often resent that their leaders – many of them holding office without being democratically elected – obtain enormous personal financial benefits from the companies, but do not share these benefits with their subjects. To protect their privileges against the general resentment of the population, dictators in oil-rich countries tend to run very repressive regimes.

Not surprisingly, conflicts about the sharing of the benefits from petroleum have sometimes led to civil wars of varying levels of intensity, as well as military coups whose perpetrators aimed to obtain high office to personally gain control of oil revenue stream provided by the resident oil company, regardless of whether that oil company was under direct control of the local government or was a foreign company operating a concession. Among the more well-known civil wars centered around oil and its benefits was the Biafra war of secession from Nigeria (1967–1970), which caused an estimated two million civilian deaths from starvation in the Nigerian district of Biafra, a region that encompasses the Niger River delta with its rich oil fields, and ended with the defeat of the rebels. A second tragic example was the bloody civil war that led to the UN-brokered secession of South Sudan from Sudan in 2011. Unfortunately, the creation of the Republic of South Sudan did not end the fighting there, and currently two factions are carrying out a very bloody war in South Sudan for the control of this new country and its lucrative oil fields.

CONCLUSION

Coal and petroleum represent stored sunlight energy that arrived on our planet for many millions of years, all the while being harvested by photosynthetic plants and microorganisms. The discovery of coal and oil deposits and the development of methods to exploit these resources on a large scale have led to a tremendous increase in the rate of overall technological and scientific developments of human society in the last couple hundred years. Some of these advances have, in turn, caused large improvements in the quality and length of human life and an explosive population growth. Such advances have often been at the expense of many plant species, whose habitats have been encroached on by man-made structures such as buildings, roads, and arable land cleared by humans of all natural vegetation and reserved for the growth of a few select plant species.

Other technological advances, however, have increased the ability of humans to wage war on one another with substantially greater lethal outcomes. As we have seen in this book, the causes of these wars have been the struggle for direct or indirect control of resources provided by a few human-favored plant species, and the resources to fight these wars are also provided, directly or indirectly, by plants both alive and dead. It is illogical to assume that human population growth could continue *ad infinitum*, and that plants and their habitats will continue to decrease. It is possible that humans and plants will reach some dynamic equilibrium on the planet. However, it is more likely that this struggle will end up in the total victory for plants and the annihilation of the human species, an outcome that will be in no small part due to plants' proclivity to cause humans to wage war on other humans as well as to provide them with the wherewithal to do so.

NOTES

1. Freeze, B. 2003. *Coal: A Human History*. Penguin Books.
2. An excellent history of oil can be found in Maugeri, L. 2006. *The Age of Oil: The Mythology, History, and Future of the World's Most Controversial Resource*. Praeger. Additional background information is also found in Black, B. C. 2012. *Crude Reality: Petroleum in World History*. Rownan and Littlefield Publishers, Inc.; and Downey, M. 2009. *Oil 101*. Wooden Table Press.
3. Medlin, L. K. 2016. Evolution of the diatoms: major steps in their evolution and a review of the supporting molecular and morphological evidence. *Phycologia* 55: 79–103; and Tirichine, L., Rastogi, A., and Bowler, C. 2017. Recent progress in diatom genomics and epigenomics. *Current Opinion in Plant Biology* 36: 46–66.
4. This is also the case for algal and plant cells and their chloroplasts.
5. See the extensive discussion of this topic in *Guns, Germs, and Steel* by Jared Diamond (W.W. Norton & Company, 1997).
6. The internal combustion engine is more efficient than steam engines in numerous ways. It uses hydrocarbons, a more concentrated form of energy than coal. These hydrocarbons are liquid, so they are much more easily stored, and can be easily transferred from one container to another and to the engine itself via pipes without the need of peoples with shovels. Trains and ships powered by steam engines needed a large number of people to handle the storage of coal and the constant coal-feeding and maintenance of the boilers. The internal combustion engine is also much more amenable to miniaturization, allowing for the development of smaller automotive devices such as planes, automobiles, motorcycles, and lawn mowers, among others.
7. The organic chemistry industry actually had its origin when people tried to use coal products to manufacture some of these products, but crude oil turned out to be a much better starting material.

8. The region containing the third major coal field under control by Germany before the war, Upper Silesia, was lost to Germany when Poland officially annexed it in 1949 following a peace treaty with East Germany.
9. Judt, T. 2005. *Postwar: A History of Europe Since 1945*. Penguin Books.
10. The United Kingdom was invited to join, but declined, since it did not want to give up its national sovereign right to control its own resources. It eventually joined the EEC, but in 2016 a plebiscite to leave the EU was approved by a slight majority of Britons.
11. Inner Manchuria and Outer Manchuria, the latter controlled by Russia by the end of the 19th century, constitute the region generally known in the West as Manchuria. This name was given to the region by Japan and is not usually used by Chinese sources.
12. The jet engine was invented prior to World War II, but its development was greatly accelerated by the war efforts.
13. TNT was originally made in 1863 by the German chemist Julius Wilbrand (Couteur P. and Burreson, J. 2003. *Napoleon's Buttons*. Jeremy P. Tarcher/Penguin) in a process that can be described in a simplified way as mixing toluene with nitric acid:

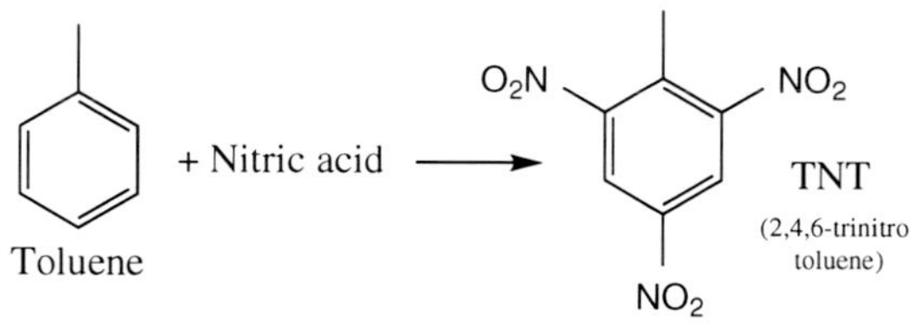

    Toluene + Nitric acid → TNT (2,4,6-trinitro toluene)

    O_2N NO_2 NO_2

    However, the explosive properties of TNT were not discovered until 1891. Toluene is a chemical that can be obtained from natural wood distillation, but at the time that Wilbrand first synthesized it, toluene was much more cheaply obtained from coal. Today, toluene is made from petroleum distillation.
14. Churchill himself pushed for his government to become a majority owner of BP in 1914, in effect nationalizing it, because of the strategic importance of petroleum.
15. Israel attacked Egypt first. Following the Israeli attack, the French and British governments announced that they were sending troops to the area to separate the combatants and impose a truce.
16. Statistics are difficult to come by, but a 2010 report from the British Broadcasting Company (Olivia Lang, *The danger of mining around the world*. BBC News, www.bbc.com/news/world-latin-america-11533349, 2010) estimated that roughly 12,000 people are killed annually around the world in accidents related to coal mining, 80% of them in China. An analysis of official Chinese reports (Wang, M.X., et al. 2011. Analysis of national coal-mining accident data in China, 2001–2008. *Public Health Reports* 126: 270–275) showed that between 2001 and 2008, annual deaths ranged from 1,086 to 4,899, but it also indicated that these records were incomplete and that it was likely that accidents and death rates were grossly underreported.
17. In the Ludlow Massacre, Colorado National Guards as well as Colorado Fuel and Iron Company guards opened fire on a tent city of striking miners demanding better working conditions from the coal mine owner, John D. Rockefeller. Some 20 people were killed, including women and children.
18. While the miners were nursing many grievances, the most recent one was that early in that August a sheriff by the name of Sid Hatfield, a union sympathizer, was murdered on the steps of a courthouse in McDowell County by members of the security force of the Baldwin-Felts Detective Agency. This agency was employed by various coal mine owners around the country to intimidate miners and break their strikes, and its agents were responsible for committing many murders that usually went unpunished.

Appendix
Chemical Notations

Molecules are made from several atoms linked together. Synonyms for molecules are compounds, chemical structures, or, in short, chemicals. Organic compounds are defined as compounds that are made in living organisms, and their hallmark is the presence of one or more carbon atoms. Besides carbon, whose chemical notation is C, other common atoms in organic molecules are hydrogen (H), nitrogen (N), oxygen (O), sulfur (S), and phosphorus (P). The atoms are linked together by covalent bonds, which are the strongest chemical bond, and are denoted by a straight line (—). Carbon can have four covalent bonds, hydrogen can have only one bond, nitrogen can have up to four bonds, oxygen two, sulfur four, and phosphorus four.

There are many ways to depict the structure of organic compounds. In Figure A.1, four ways of depicting the compound acetone, whose formula is C_3H_6O, are shown. On the left, all the atoms and the bonds are shown in full. However, organic molecules often have many hydrogen atoms, and depicting all of them with their bonds causes clutter, so often these hydrogens will not be depicted and instead the number of hydrogen atoms linked to each atom of another kind are shown, as depicted in the structure second from left in the diagram below.

To simplify even further, hydrogen atoms are not shown at all (third from left), since their numbers could be deduced by the reader from the known rules (for example, a carbon that is linked by one covalent bond to another carbon is also linked to three hydrogen atoms, even when these hydrogen atoms are not shown). However, on a few occasions some hydrogen atoms linked to carbon will selectively be shown if doing so provides additional information or improves clarity. Finally, since there are also many carbon atoms in organic compounds, the carbons themselves may not be shown but only the covalent bonds between them, as shown in the structure on the right side of the diagram.

The same rules apply to more complicated organic molecules and with more type of atoms. For example, the amino acid phenylalanine (abbreviated as Phe) can be depicted in full as shown on the left in Figure A.2, or in the abbreviated, simpler to look at form on the right:

However, Phe presents one more complication that is often encountered in organic molecules and that needs to be explained here (Figure A.3). Phe has one carbon that is

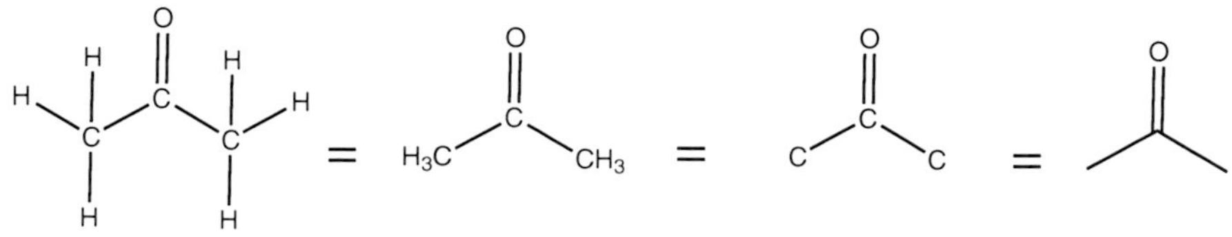

FIGURE A.1 Alternative ways to depict the structure of acetone.

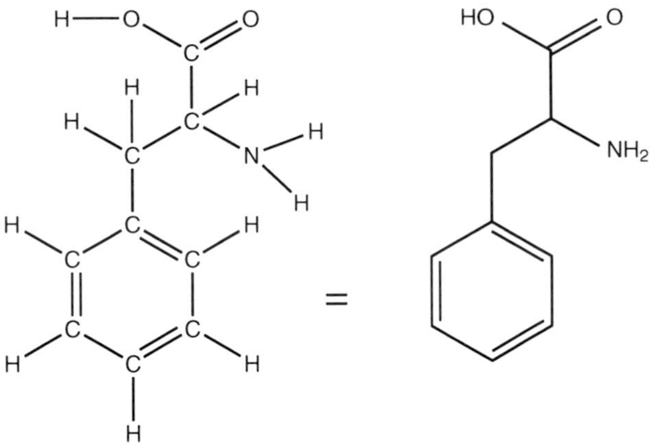

FIGURE A.2 Alternative depictions of phenylalanine.

linked to four separate entities: one hydrogen (green circle); one nitrogen that is linked to two hydrogens (pink circle); one carbon that is linked to two oxygen atoms, one of which is linked to one hydrogen atom (orange circle); and one carbon which is part of a large constellation of atoms (blue circle). It turns out there are two ways to link these four groups of atoms to the central carbon atom. To visualize it in three dimensions, think of the two carbons that are linked to this central carbon as residing on the plane of the paper (or screen), whereas the nitrogen and hydrogen atoms are on either side of the plane, one below it and the other above it. A link to an atom below the plane is depicted with an empty arrowhead (or sometimes with a broken arrowhead), and a link to an atom above the plane is shown as a solid arrowhead. Thus, there are in fact two types of Phe molecules that are mirror images of each other, one in which the hydrogen is below the plane and the nitrogen is above the plane (Figure A.3, left), and the other where the two atoms switch position (Figure A.3, right). These two

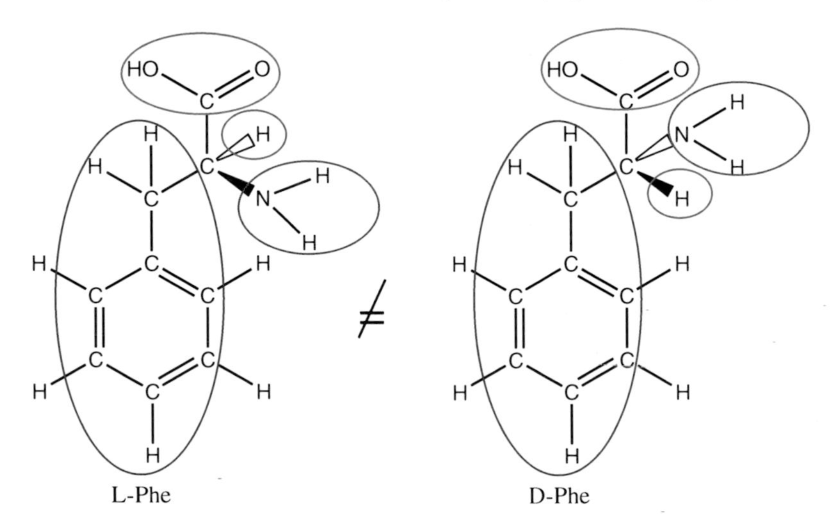

FIGURE A.3 L and D stereoisomers of Phe.

compounds are called enantiomers (or sometimes, imprecisely, as stereoisomers), in this case distinguished by the designations D and L. The Phe that is a component of most proteins is L-Phe, and in general living organisms have enzymes that reliably make and use one type of stereoisomer and not the other.

A carbon that is bound to four different groups is called a "chiral center." Some molecules (but not Phe) have more than one chiral center.

Scientific Glossary

Amino acids: A type of acid molecule that contains a nitrogen atom. There are many types of amino acids found in nature, but all living organisms use the same group of 20 amino acids as the main building blocks of their **proteins**.

Alkaloid: Organic chemicals that contain one or more nitrogen atoms, often as part of ring of atoms. Because of the nitrogen atoms, when alkaloids are dissolved in water the solution becomes basic, or alkaline (i.e. pH > 7). This property is the basis for calling such compounds "alkaloid."

Angiosperms: The evolutionary related group of plants that have flowers and produce seeds.

Cell: A biological cell is a structure that consists of outer membranes made of water-repelling (hydrophobic) material that limits the movement of compounds in and out of the space confined within. Inside, bacterial, plant, and animal cells all have genetic material and various other chemical compounds that interact with each other and create a unique chemical environment that is different from what is on the outside.

Chemical compound: A molecule that has more than one atom, linked to each other. Often just called a chemical, a compound, or a molecule.

Chromosome: A structure inside the living cell that contains a long linear DNA molecule (which carries many genes). Each cell typically has a number of chromosomes, each carrying a fraction of the genes that the cell carries. For example, a female human has 23 distinct chromosomes of unequal length, and each chromosome is present in two copies in each cell (since humans are **diploid**). When completely extended, each chromosome is many times longer that the length of the cell. In order to fit inside the cell, the chromosome is folded in a compact manner. Each cell in the body of a multicellular organism, such as an animal or a plant, carries the same complement of chromosomes.

Colloidal: The property of large molecules, too big and often too hydrophobic to be completely dissolved in water, to nonetheless stay suspended in aqueous solution, a state called an emulsion. Due to the large size of particles in an emulsion, they reflect light, giving the emulsion white color unless the particles absorb light of specific wavelength in the visible range. In the latter case, the color of the emulsion will be that of the reflected light not absorbed by the particles.

Diploid: Having two sets of identical chromosomes. Most animals and many plant species are diploid. Quite a few plant species are **tetraploid**, having four sets of identical chromosomes, and some **hexaploid** (six sets), **octaploid** (eight sets), or have higher ploidy.

Conspecific: A member of the same species as another individual with which it is compared to. Sometimes called congener.

DNA: The chemical structure that carries the genetic material. DNA (short for deoxyribonucleic acid) is a long polymer made up of "ribonucleotide" subunits linked together in tandem. Each ribonucleotide has three components – a phosphate group, a sugar (deoxyribose), and one of four bases, abbreviated as A, C, G, and T.

Ecology: The scientific field of study of how living organisms interact with other living organisms and with the environment.

Embryo: The early developmental stages of a multicellular living organism after female and male **gametes** (**egg** and **sperm** cells, respectively, which are cells that contain half the set of chromosomes present in cells of the body of the individuals that produce such gametes) fuse together into a single cell, called a **zygote**, that then begins to divide and develop into its final shape and size.

Fermentation: Originally, the biochemical process of breaking down sugar in the cell to alcohol and carbon dioxide, but often applied to other metabolic processes that break down other metabolites.

Fertilization: The fusion of the egg and sperm cells. In animal biology, the term is also often used to describe the process of delivering the sperm to the egg, although in plant biology that latter process is often called **pollination**.

Fitness: Biological fitness is defined as the relative success (compared to conspecifics) of passing one's genes to the next generation, i.e., the number of progeny one begets.

Gene: A segment of a DNA molecule that contains the information for machinery in the cell to make a specific molecule, usually a protein (via an RNA intermediate) but sometime just an RNA molecule, that participates in chemical reactions in the cell.

Genome: The total complement of genes that a cell has on all of its chromosomes.

Gymnosperms: The evolutionary related group of plants that have cones, not flowers, as their reproductive organ and produce seeds that are not covered by additional plant tissue. Coniferous trees such as pines are members of the gymnosperm group.

Meiosis: The process of generating gametes from germ cells (see **Sexual reproduction**).

Metabolite: A chemical compound that is produced inside a living cell by changing the composition of precursor chemical compound via a chemical reaction, a process called **metabolism**.

Neurotransmitter: A chemical that is released at the end of one nerve cell at the synapse. Once released, the many molecules of this chemical then diffuse across the synapse to the adjacent nerve cell whereby they bind to **neuroreceptors**, which are made of proteins.

Organic chemicals: Chemical compounds that are made in living organisms and have at least one carbon atom but most often several carbon atoms linked to each other. Organic molecules also often have hydrogen, oxygen, and/or nitrogen, and less often sulfur, phosphorus, and a few other elements.

pH: A measure of the concentration of hydrogen cations (H^+) in aqueous solutions, using a negative log scale. pH values <7 indicate more H^+ cations than is found in distilled water, and thus indicate acidity, and pH values >7 indicate fewer H^+ cations than is found in distilled water, and thus indicate basic or alkaline solution.

Phloem: A vascular system inside plants where a water solution containing various metabolites such as sugars and amino acids flows from one part of the plant to another.

Photosynthesis: A biochemical process in which cells use energy captured from sunlight to split water (H_2O) by taking two electrons from this molecule and using these electrons to reduce the carbon present in gaseous carbon dioxide (CO_2) to build organic molecules based on a carbon skeleton. Photosynthesis in plants occurs in a special structure inside the cell called a chloroplast. The oxygen atom that is liberated from the split water molecule combines with another oxygen atom to form O_2, a gas which we breath and use to carry out our own metabolic processes of generating energy.

Pollination: The transfer of pollen grain from the anthers of a flower to the stigma of a flower (when the same flower, the process is called "**selfing**." Pollen transfer can occur via the action of animals ("**pollinators**") or abiotic factors such as wind. Once pollen land on the stigma, the pollen grain grows inside the stigma toward the ovule, where the egg cell is, and delivers the sperm cell that is inside the pollen grain cell.

Population: A group of living organisms of the same species that live in proximity to each other and therefore sometimes breed with each other.

Population genetics: The study of changes in the genetic information of individuals in a population and of the entire group.

Protein: A long linear polymer of amino acids connected to each other through carbon–nitrogen linkage. Some proteins have only a few scores of amino acids and some have more than a thousand, but the majority of proteins in the cell of any living organisms contain a few hundred amino acids. Proteins typically do not stay as long, straight chains but fold into tight, compact structures that have unique properties such as surfaces that bind specific metabolites or other proteins. Proteins often serve as catalysts ("enzymes") that speed up metabolic reactions in the cell.

Psychoactive chemicals: Chemicals that, once they come into contact with the nervous system of an animal, change the behavior of the organism.

RNA: A molecule similar to DNA, but containing ribose instead of deoxyribose (therefore the acronym is ribonucleic acid) and the base U instead of T (the other three bases, A, C, and G, are the same as in DNA).

Sexual reproduction: A method of generating a new living organism by the fusion of two cells (called **gametes**), each originating from two different individuals, or, in the case of hermaphroditic organisms, from two different germ cell lines within the same individual. With plants and animals, the two parents (or germ cell lines) involved in sexual reproduction are called males and females, and the gametes that come from them are called sperm and egg, respectively. While the cells of each parent have two sets of genes (i.e., two copies, or alleles, of each gene, with the two copies being either identical, or not), the single set of genes present in the gamete that is generated by the parent is assembled by taking half the time the gene copy from its set #1 and the other half the gene version from its set #2. Since the particular assembly of the gene set in each gamete is different every time a gamete is formed (basically, which version of each gene is included in the gamete is determined by luck), each gamete is genetically unique, and the progeny each progeny produced by the fusion of gametes has a unique gene combination that is not identical even to its siblings. The process of generating gametes from germ cells is called **meiosis**.

Sociobiology: The field of study exploring how genetic and environmental factors, including the behavior of conspecifics, jointly affect individual living organisms.

Synapse: The place, sometimes also called a junction, where the ends of two nerve cells meet. The signal from one nerve cells is transmitted to the other via the synapse (pl. synapses).

Vegetative reproduction: Any number of ways in which a new organism is created from an existing one without the fusion of an egg and a sperm cell. Also called asexual reproduction or cloning. Common in plants and microorganisms, but also present in animals.

Vine: A plant that has weak stems so that they cannot stand upward on their own. Vines typically grow in the vicinity of trees and bushes and use their stems for support.

Xylem: The vascular system of plants that convey water from the roots to the aerial parts of the plant.

Index